Elementary Materials Science

William F. Hosford

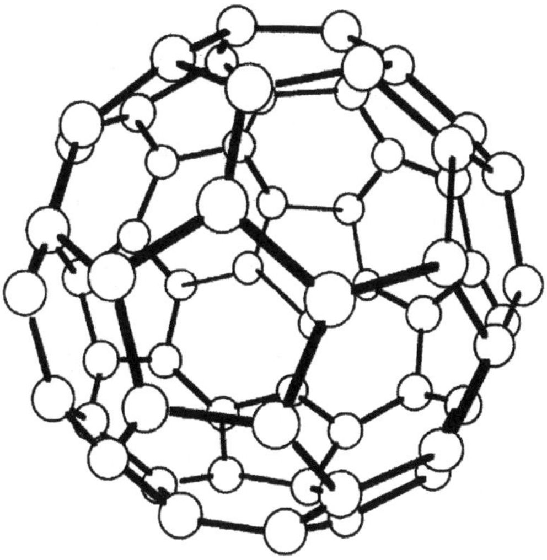

The Materials Information Society

ASM International®
Materials Park, Ohio 44073-0002
www.asminternational.org

First printing, August 2013

Prepared under the direction of the ASM International Technical Book Committee (2012–2013), Bradley J. Diak, Chair.

ASM International staff who worked on this project include Scott Henry, Senior Manager, Content Development and Publishing; Karen Marken, Senior Managing Editor; Amy Nolan, Content Developer; Sue Sellers, Editorial Assistant; Madrid Tramble, Manager of Production; and Diane Whitelaw, Production Coordinator.

Library of Congress Control Number: 2013931785

ISBN-13: 978-1-62708-002-6

ISBN-10: 0-62708-002-3

SAN: 204-7586

ASM International®
Materials Park, OH 44073-0002

www.asminternational.org

Printed in the United States of America

Contents

Preface

This textbook, which covers the subject of materials science with very few equations, is designed for students who are interested in materials science. It is assumed they already have had a course in chemistry. A prior course in physics is not necessary, and the use of mathematics is limited to algebra. This text also will be useful to nontechnical professionals in the materials industry.

The book first introduces materials science through the discussion of the elements, bonding, crystal structures, and amorphous (noncrystalline) materials; then presents the properties of phase relations, mechanical behavior, electrical behavior, and magnetic behavior. There is one chapter each devoted to the following materials: nonferrous metals, ferrous metals, ceramics, polymers, composites, and wood. To round out this important basic volume on a growing interdisciplinary scientific field are chapters on corrosion, forming and shaping, and recycling. There are 14 chapters. Chapters 3 to 7, and 9 will likely require more time to cover than chapters 1, 2, 8, 10, and 12. Chapters 11, 13, and 14 can be covered in less time than the others. Chapters end with a Note of Interest and Exercises—all pertaining to the chapters' topics of discussion. The Notes of Interest engage readers with fascinating bits of information about notable people, events, and developments; the Exercises allow students to explore the world of materials in a pragmatic, hands-on manner.

CHAPTER 1

Chemistry Basics

MATERIALS are so important to civilization that the terms *stone age*, *bronze age*, and *iron age* have been used to describe periods of history. Perhaps the present day will be called the *silicon age*. Without steel, modern inventions such as the automobile, the bicycle, and railroads would be impossible. Without aluminum, there would be no airplanes. Without silicon, there would be no electronic devices like computers, digital cameras, or cell phones.

Advancements in materials have been responsible for most of today's prosperity. Two centuries ago the most important materials were wood and clay. Glass and steel, wrought iron, copper, silver, and pewter were available, but were very expensive. There were no plastics, processed aluminum, modern rubber, and semiconductors. In the mid 1800's advances in steelmaking made the production of steel much less expensive. The technique for economically producing aluminum was developed in the late 1800's. The first plastics appeared in the early 1900's, but only a few were available until the mid 1900's. The development of semiconductors in the last half of the twentieth century made possible the wide variety of electronic products in common use today.

Elements

Before delving into a discussion of materials, it will be helpful to take a look at the particles that materials are composed of, and the chemistry and forces at work behind them. An overview of chemistry is, in fact, necessary to arrive at an understanding of why and how materials have the characteristics they do.

The periodic table of the elements (Fig. 1.1) allows one to quickly see which chemical elements are similar to others. The elements are arranged in order of increasing atomic number.

The table includes periods (typically horizontal) and groups (typically vertical). Elements in groups share some similar properties.

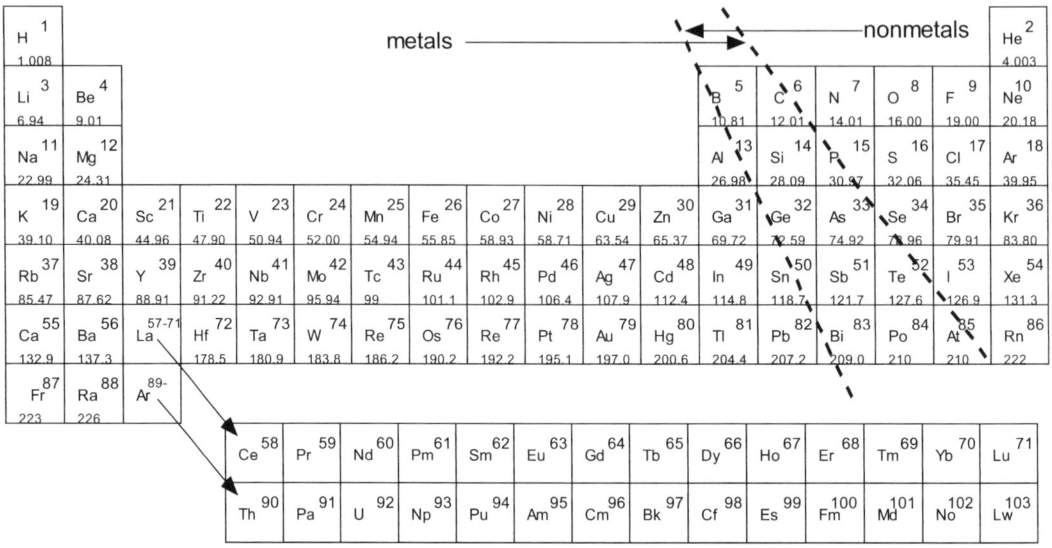

Fig. 1.1 The periodic table of the elements. The atomic numbers are in the upper right corners and the atomic wts (in amu) at the bottom.

Fundamental Particles

All solid materials are composed of atoms, ions, or molecules. *Atoms* are the basic unit of chemical elements. *Ions* are simply atoms or groups of atoms that have lost or gained one or more electrons and carry an electric charge. *Molecules* are groups of atoms chemically bonded to each other.

Bonding

The study of materials requires an understanding of the forces that hold atoms together as solids. There are four types of bonds: metallic, ionic, covalent, and van der Waals. In metallic bonding, individual atoms lose their valence electrons, which forms a negatively charged "electron gas." This electron gas holds together the positively charged atoms. In ionic solids there is a transfer of valence electrons. Anions (metallic ions) have lost valence electrons and become positively charged, while cations (nonmetal ions) are atoms that have accepted extra electrons and become negatively charged. Electrostatic attraction bonds anions to cations. Figures 1.2(a) and (b) illustrate metallic and ionic bonding, respectively. In covalent bonding (Fig. 1.2c), valence electrons are shared so that outer shells can be complete. For example, if a carbon atom with four valence electrons shares one electron with each of four neighboring carbon atoms, each carbon atom will have eight outer electrons. Often bonding between atoms is partially ionic and partially covalent in character. Van der Waals (Fig. 1.2d) bonding is much weaker than metallic, ionic, and covalent bonding. It arises from the electrostatic attraction of molecules. Asymmetric molecules are likely to have dipole moments (meaning the ends have different charges), and bonding results from the attraction between the dipoles. Even symmetric molecules and atoms have statistical dipoles, which results in even weaker bonding. Hydrogen atoms covalently bonded in water molecules create strong dipoles. The term *hydrogen bonding* refers to van der Waals bonding resulting from these dipoles. Figure 1.3 illustrates this schematically.

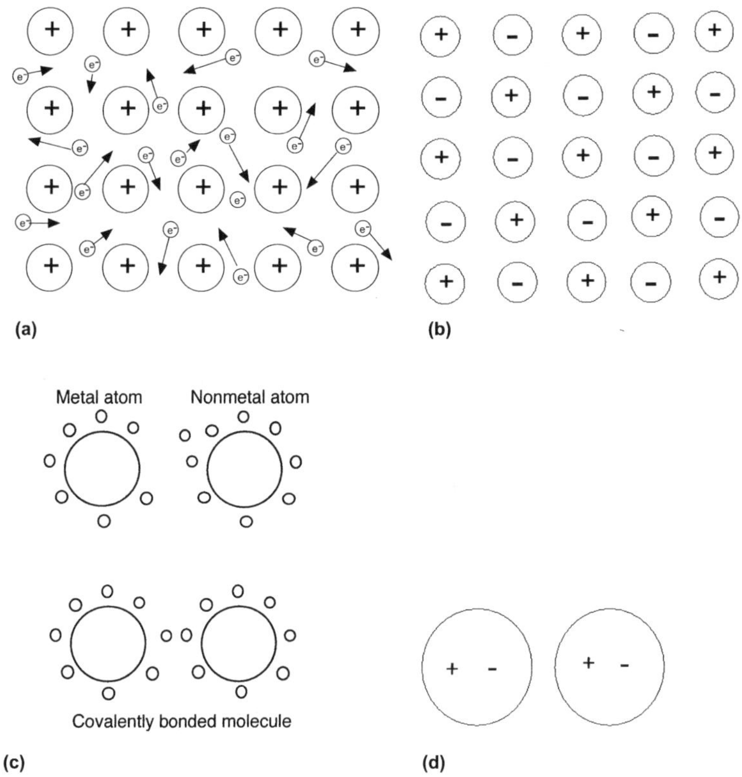

Fig. 1.2 (a) Metallic bonding results from attraction of positive atoms to electron gas. (b) Ionic bonding results from mutual attraction of anions and cations. (c) Covalent bonding involves sharing of electrons to complete outer shells. (d) Van der Waals bonds are the result of statistical dipoles.

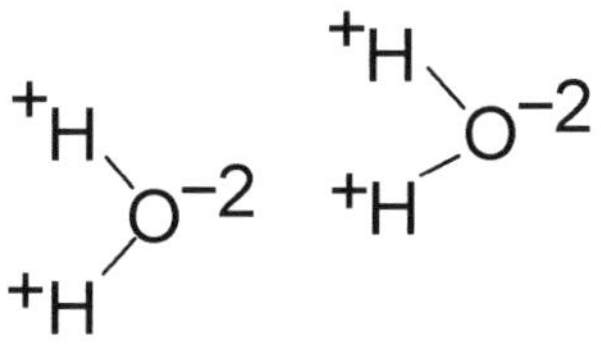

Fig. 1.3 Hydrogen bond between adjacent water molecules resulting from the dipole of the hydrogen atom

Table 1.1 Bond strengths

Bond type	Energy (kJ/mol)
Van der Waals	0.0–10
Hydrogen	10–40
Ionic	50–1000
Covalent	200–1000
Metallic	50–1000

Source: Ref 1.1

Table 1.1 gives the approximate ranges of the strengths of the different types of bonds.

It is useful to think of atoms and ions as hard spheres. In solids and liquids, these hard spheres are in contact with each other.

Crystal Structures

Solids can be classified into those that are crystalline and those that are not. A *crystal structure* is one that repeats itself over long distances. Most solids are crystal structures, or crystalline. The atoms, ions, or molecules are arranged in repeating three-dimensional patterns or lattices. The smallest characteristic piece of the repeating pattern is called the *unit cell.* In metal, crystal atoms are in contact with each other. Most crystal structures are either the densest way of packing spheres or almost the densest. Most metals crystallize in one of three crystal structures.

Face-Centered Cubic Crystal Structure

One of these is called *face-centered cubic* (fcc) because the unit cell (Fig. 1.4) is cubic with an atom on each corner of a cube and one in the center of each face.

The atoms on each corner of the unit cell are shared with eight other unit cells, so the number of corner atoms is (8 corners) (1/8 atom per corner) = 1 corner atom. There are six faces, and atoms in the centers of each face are shared with one other unit cell, so there are (6 face-centering atoms) (1/2 atom per face) = 3 face-centering atoms. Thus there are four atoms in the unit cell. In an fcc crystal, each atom has 12 nearest neighbors. The face-centered cubic structure is one of the most densely ways of packing spheres. Atoms are arranged in close-packed hexagonal planes as shown in Fig. 1.5. These planes are stacked so that the atoms in one plane are nested into the valleys of the planes below. There are two sets of valleys. In the face-centered cubic structure, the planes are stacked so that the third plane is not directly over the first. Aluminum, copper, nickel, lead, gold, silver, platinum, as well as iron at elevated temperatures, crystallize in a face-centered cubic crystal structure.

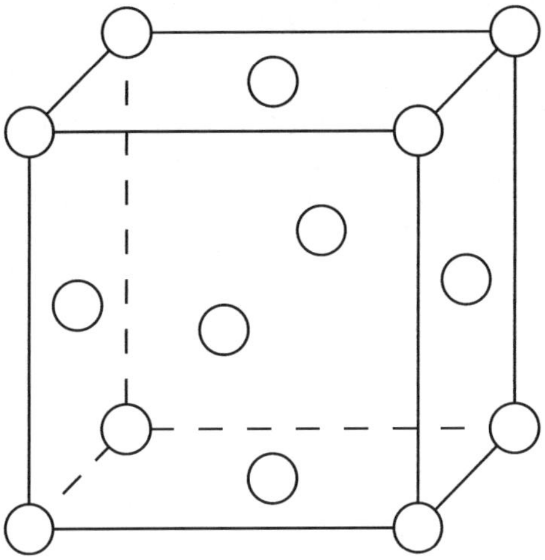

Fig. 1.4 The unit cell of a face-centered cubic structure

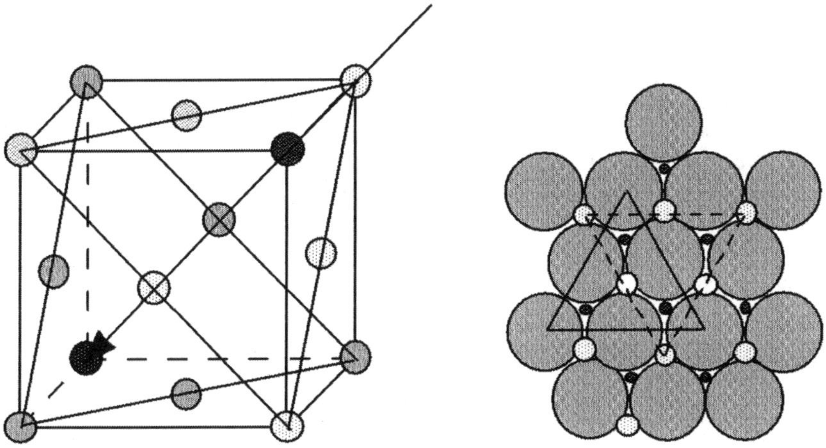

Fig. 1.5 Close packed planes in the face-centered cubic structure

Body-Centered Cubic Crystal Structure

Another common crystal structure of metals is called *body-centered cubic* (bcc) because the unit cell is cubic with an atom on each corner of a cube and one in the middle of the cube (Fig. 1.6). Each of the eight corner atoms is shared by eight unit cells so there are two atoms per unit cell (one corner atom and one body-centering atom). Among the metals having a body-centered cubic crystal structure are chromium, tantalum, vanadium, niobium, tungsten, molybdenum, iron at room temperature, and both titanium and zirconium at elevated temperatures.

Hexagonal Close Packed Crystal Structure

The third common crystal structure of metals is called *hexagonal close packed* (hcp) because the basic structure is a hexagonal prism (Fig. 1.7). The prism is composed of three layers of atoms, each arranged in a hexagonal pattern with an atom in the center of the hexagon. Only three

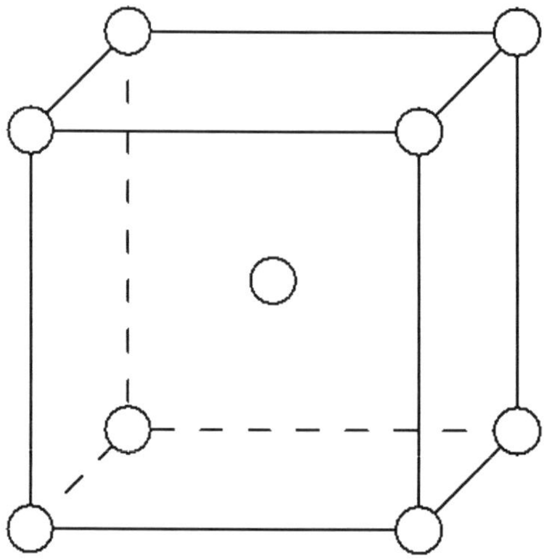

Fig. 1.6 The unit cell of the body-centered cubic structure

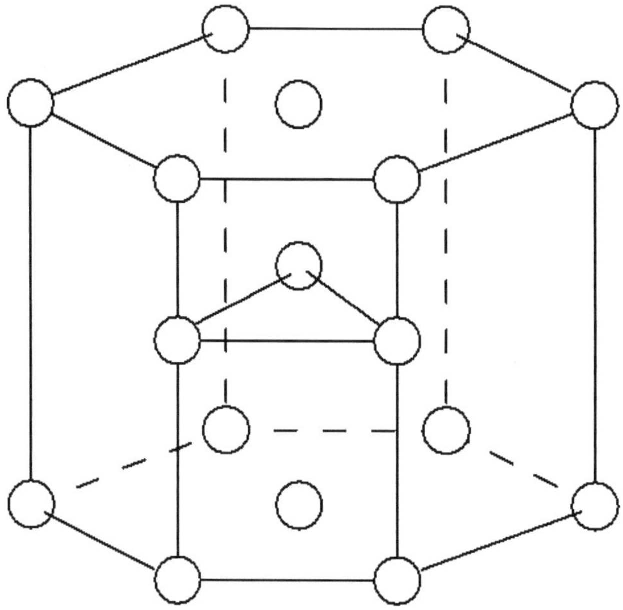

Fig. 1.7 The basic hexagonal cell of the hexagonal close packed structure

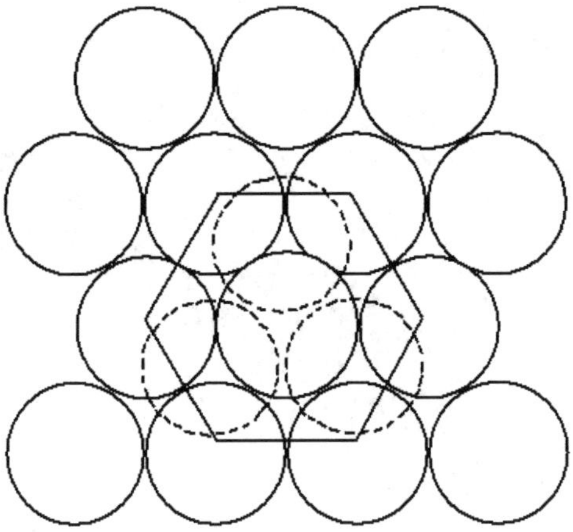

Fig. 1.8 Stacking of close packed planes in the hexagonal close packed structure

atoms of the middle plane are in the basic cell. There are two atoms in the top and bottom that are shared by two prisms, and 12 corner atoms shared by six prisms, so there are a total of 3 + 1 + 2 = 6 atoms per prism. The structure consists of stacked hexagonal planes, as shown from above, in Fig. 1.8.

Among the metals that crystallize in the hexagonal close packed structure are beryllium, magnesium, zinc, cadmium, and both titanium and zirconium at room temperature.

Diamond Cubic Crystal Structure

Aside from the three crystal structures (fcc, bcc, hcp), certain elements, including carbon, silicon, and germanium, crystallize in another crystal structure, the diamond cubic crystal structure, which has a repeating pattern of eight atoms.

The periodic table (Fig. 1.9) shows that elements with the same crystal structure are clustered together.

Sodium Chloride Crystal Structure

There are many different crystal structures of compounds. Figure 1.10 shows the structure of sodium chloride. The chlorine ions are arranged in a face-centered cubic arrangement, with the sodium ions occupying the locations where they are in contact with four chlorine ions. This structure also can be thought of as a face-centered cubic arrangement and the sodium ions with chlorine ions occupying the locations where they are in contact with four sodium ions.

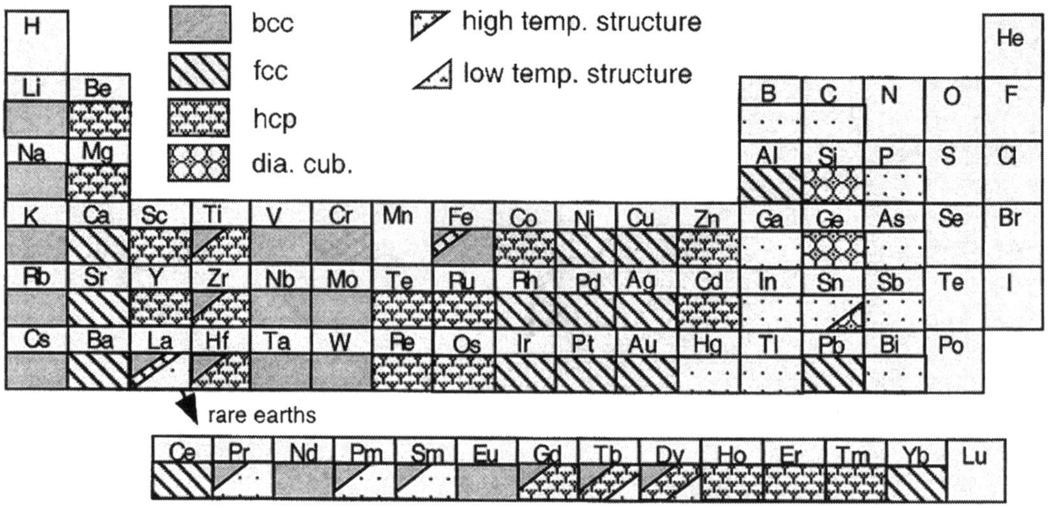

Fig. 1.9 The periodic table. Note that neighboring elements share similar crystal structures. Source: Ref 1.1

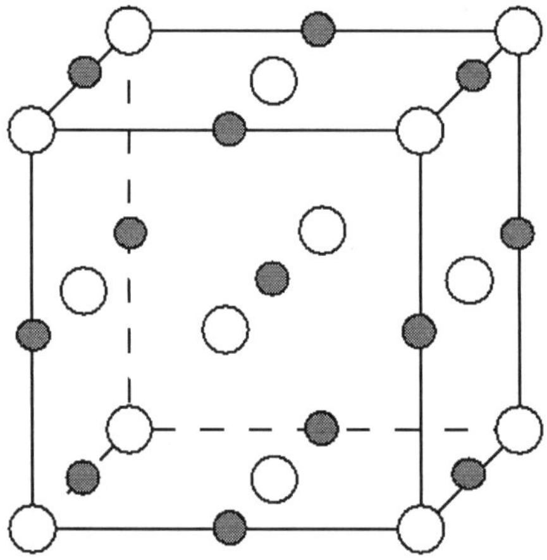

Fig. 1.10 The structure of sodium chloride

Other compounds that crystallize with the sodium chloride structure include:

barium oxide (BaO)
barium selenide (BaSe)
barium sulfide (BaS)
barium telluride (BaTe)
cadmium oxide (CdO)
calcium oxide (CaO)
calcium sulfide (CaS)
cobalt oxide (CoO)
ferrous oxide (FeO)
lead sulfide (PbS)

lithium bromide (LiBr)
lithium chloride (LiCl)
lithium fluoride (LiF)
lithium hydride (LiH)
lithium iodide (LiI)
magnesium oxide (MgO)
magnesium sulfide (MgS)
manganese oxide (MnO)
manganese sulfide (MnS)
nickel oxide (NiO)
potassium bromide (KBr)
potassium chloride (KCl)
potassium fluoride (KF)
potassium hydride (KH)
potassium iodide (KI)
silver bromide (AgBr)
silver chloride (AgCl)
sodium iodide (NaI)
tantalum carbide (TaC)
tantalum oxide (TaO)
titanium carbide (TiC)
titanium nitride (TiN)
titanium oxide (TiO)
vanadium carbide (VC)
zirconium carbide (ZrC)
zirconium oxide (ZrO)

There are many other crystal structures of compounds. In all compounds, the anions and cations are in contact. In almost every case, the structure is the one that maximizes the density of the packing of spheres of different sizes.

Amorphous (Noncrystalline) Materials

Noncrystalline solids are said to be *amorphous*. An amorphous material has a structure like that of a liquid. Atoms, ions, or molecules are in contact with each other as in a crystal. Often there is short-range order, so that each atom, ion, or molecule has nearly the same surroundings as others, but there is no repeating pattern over long distances. Glass and most plastics are amorphous. In silicate glasses, silicon atoms are surrounded by four oxygen atoms forming tetrahedra. Oxygen atoms are shared by two tetrahedra as sketched in Fig. 1.11.

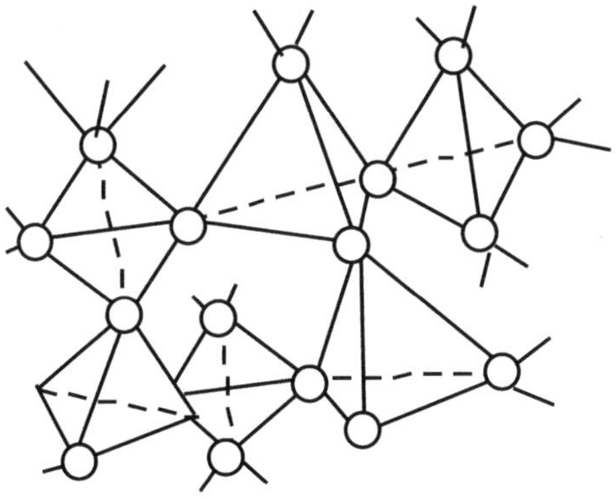

Fig. 1.11 Structure of a silicate glass consists of tetrahedra with silicon atoms in the centers and oxygen atoms on the corners. Source: Ref 1.2

NOTE OF INTEREST

The seven metals known to antiquity are iron, copper, silver, tin, gold, mercury, and lead. Those that occurred in the native (metallic) state are gold, silver, copper, and mercury. Metallic iron in the form of meteorites also was probably known early. Tin and lead are not found in a native state, but they probably were first reduced accidentally in a fire. The first metallic material to be used for tools or weapons likely was bronze, an alloy of copper and tin.

Exercises

1. **Forming hexagonal close packed and face-centered cubic crystal structures**

   Construct four rafts of balls arranged in hexagonal patterns with at least 10 balls in each raft. Then stack the rafts in different patterns. Try to stack them so they form a hexagonal close packed crystal; restack them so they form a face-centered cubic crystal.

   If the third raft is placed directly over the first raft, the result is hexagonal close packing. If it is not, but the fourth raft is placed directly over the first raft, the result is face-centered cubic.

2. **Comparing fcc and bcc crystal structures**

   Construct a face-centered cubic unit cell and a body-centered cubic unit cell. Then compare the sizes of the largest holes in each. Which would accommodate a small carbon atom more easily? Next, try to see which has the larger passages between holes. Which would allow the motion of small atoms more easily?

   The face-centered cubic cell has the larger hole which allows it to have the greater solubility for carbon. However, the body-centered cubic cell has larger passages between holes so the rates of diffusion in the body-centered cubic lattice are greater.

REFERENCES

1.1 W.F. Hosford, *Physical Metallurgy*, 2nd ed., CRC Press, 2010

1.2 W.F. Hosford, *Materials for Engineers*, Cambridge U. Press, 2008

SELECTED REFERENCES

- R.A. Flinn and P.T. Trojan, *Engineering Materials and their Applications*, 4th ed. Houghton Mifflin, 1990.

CHAPTER 2

Phase Relations

Phases

A DISTINCT STATE of aggregation of matter is called a *phase*. A phase may be a solid, a liquid, or a gas. A phase may be a pure material or a solution of several components. A solid phase is either crystalline or amorphous (noncrystalline). A crystalline phase has a characteristic crystal structure and a definite composition range. A physical system may contain more than one phase. A glass of root beer with ice contains three phases: solid ice, a water-based solution, and bubbles of carbon dioxide. A material may contain more than one solid phase, each with a different crystal structure, or one or more mutually insoluble liquid phases (for example, oil and water). All gases are mutually soluble, so only one gaseous phase is possible in a material.

Single-phase solids include brass (a solid solution of zinc in copper, with zinc atoms occupying lattice sites), sodium chloride crystals, glass, and polyethylene. Most plain-carbon steels are two-phase materials, consisting of an iron-rich solid solution containing a small amount of carbon in solution, and iron carbide.

The composition of a single phase may vary from place to place but always varies gradually without abrupt changes. In multiple phase systems, however, there are discontinuities of composition and structure at phase boundaries. The different compositions on one side of the phase boundary are in equilibrium with the compositions on the other side of the boundary. For example, the oxygen concentration changes abruptly between copper with some oxygen dissolved in it (on one side of the phase boundary), and copper oxide (on the other side).

Solid Solutions

Solid phases need not be pure materials; they also can be solutions. Crystalline solids may dissolve other elements. Solution of element B in

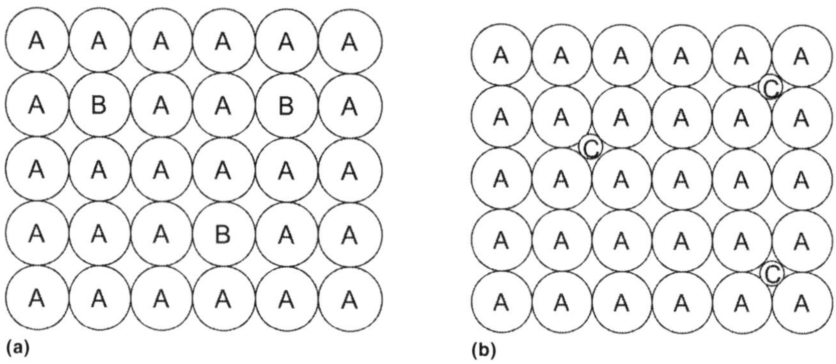

Fig. 2.1 A substitutional solid solution (a) and an interstitial solid solution (b)

crystals of A can occur by atoms of B substituting for A atoms in the crystal structure. This is called a *substitutional solid solution* and is illustrated in Fig. 2.1(a). If the atoms of element C are very much smaller than the atoms of A, they may dissolve by occupying the holes between the A atoms. This is called an *interstitial solid solution* and is illustrated in Fig. 2.1(b). Only atoms of hydrogen, boron, carbon, and nitrogen are small enough to dissolve interstitially in metals.

Phase Diagrams

Phase diagrams show the regions of stability of the various phases of a system at equilibrium. Equilibrium between two phases can be described by the limits of solubility in the two phases. A phase diagram also may be thought of as a map of permissible phases. Equilibrium between two components consists of single-phase regions and two-phase regions.

In a solid alloy of lead and tin, solid lead can dissolve tin, and solid tin can dissolve lead. Figure 2.2 is a plot of the solubility limits in the lead-tin system.

If the lead-rich solid solution is designated by α and the tin-rich solid solution by β, the fields of possible compositions can be labeled as shown in Fig. 2.3. Any composition in the $\alpha + \beta$ region between the solubility limits consists of two-phases, α and β. Both phases are saturated solid solutions. Their compositions are given by the solubility limits at that temperature.

At higher temperatures tin and lead form a liquid solution. Lead-rich compositions precipitate crystals of α on cooling, and tin-rich solid solutions precipitate β crystals on cooling. Above 185 °C (365 °F), the solubilities of tin (Sn) in the α phase and of lead (Pb) in the β phase decrease. These solubility limits are plotted in Fig. 2.4. The line from 0% Sn at 327° to 62% Sn at 183° and the line from 62% Sn at 183° to 100% Sn at 232° represent the solubilities of lead and tin, respectively, in the liquid. The line from 0% Sn at 327° to 19% Sn at 183° represents the solubility of Sn in the α phase, and the line from about 98% Sn at 183° to 100% Sn at

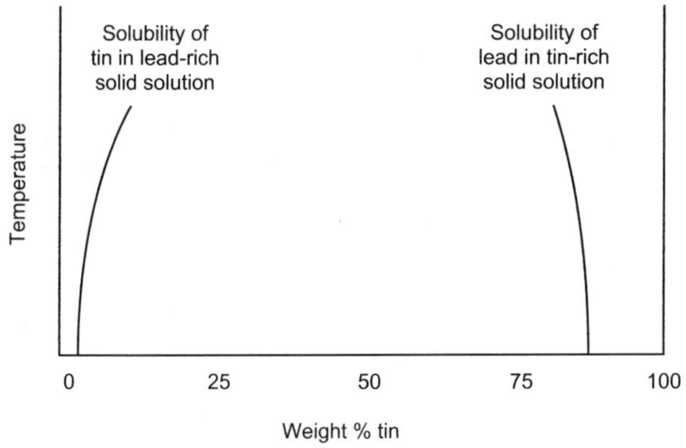

Fig. 2.2 Solubility limits in the lead-tin system

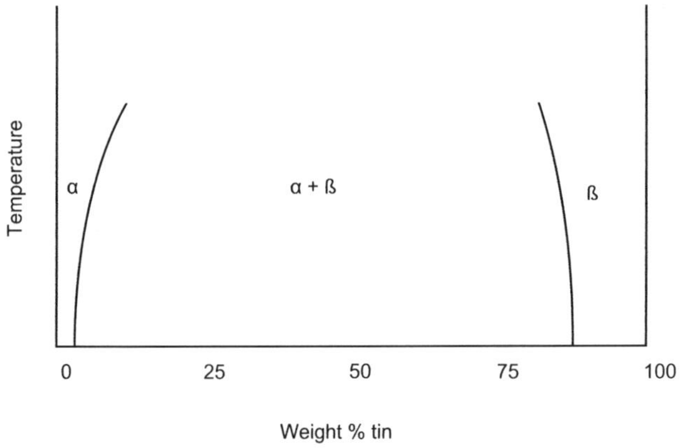

Fig. 2.3 Phase regions in the lead-tin system

232° represents the solubility of lead in the β phase. These lines complete the lead-tin *phase diagram*. This system is called *eutectic*.

The reaction L → α + β on cooling through 183° is a *eutectic reaction* that on cooling, a liquid transforms into two solid phases at the same time.

When a liquid containing 62% Sn cools, the α and β phases form a eutectic structure of alternating platelets as sketched schematically in Fig. 2.5.

In another system, copper and nickel are completely soluble in each other in the liquid state and in the solid state. Figure 2.6 shows the phase diagram of this system.

The lower line is the solubility limit of copper in the solid; the upper line is the solubility limit of copper in the liquid.

The copper-tin phase diagram (Fig. 2.7) is more complicated. Above 798 °C (1472 °F), an alloy of 22% Sn consists of two phases, liquid and α. The

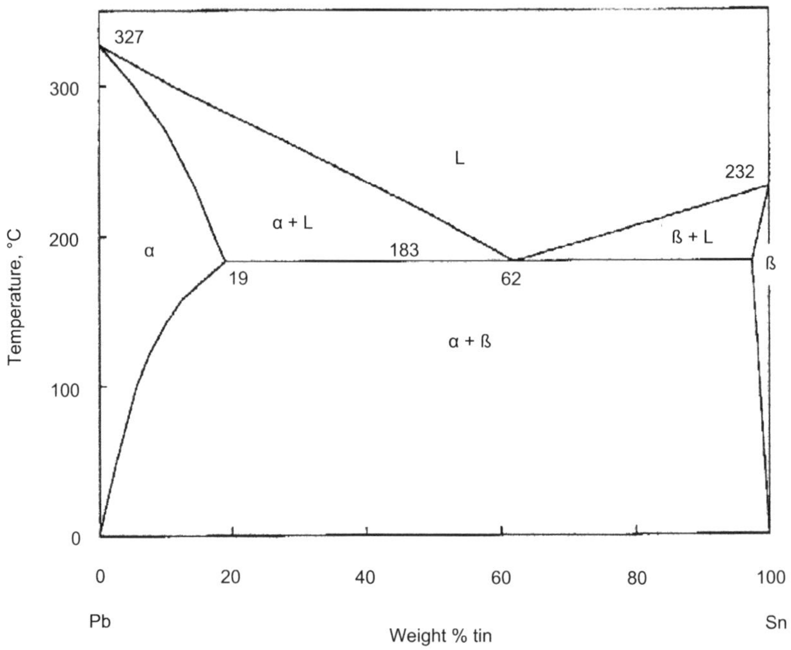

Fig. 2.4 The lead-tin phase diagram. Source: Ref 2.1

Fig. 2.5 The eutectic structure of a lead-tin alloy consisting of alternating platelets of lead-rich and tin-rich phases. Source: Ref 2.1

reaction L + α → β occurs as this alloy is cooled through 798 °C (1472 °F). This reaction, in which a solid phase and liquid phase will form a different solid phase on cooling, is called a *peritectic reaction*. Note that in the Cu-Sn diagram, there is another peritectic reaction: L + β → γ.

Another important reaction occurs in the iron-carbon phase diagram (Fig. 2.8). Above 727 °C (1340 °F) iron containing 0.77% carbon has a

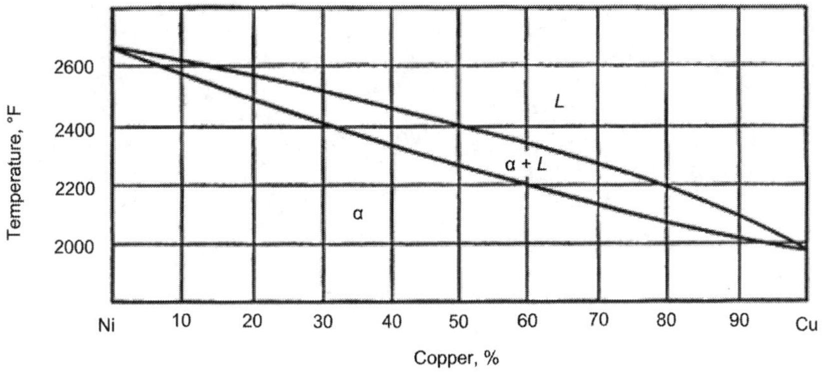

Fig. 2.6 The copper-nickel phase diagram. Source: Ref 2.2

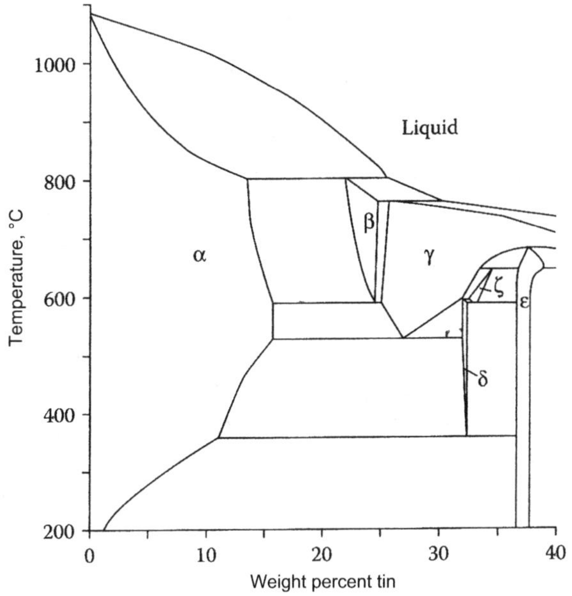

Fig. 2.7 The copper-tin phase diagram. Source: Ref 2.3

face-centered cubic structure (γ) called *austenite*. On cooling through 727 °C (1340 °F), austenite transforms to ferrite and iron carbide, $\gamma \rightarrow \alpha + Fe_3C$. This reaction of one solid transforming into two others on cooling is called a *eutectoid* reaction.

For an overall composition in a two-phase region, simple mass balance can be used to find the relative amounts of each phase. For example, at 600°C, an alloy containing 80% copper and 20 % silver consists of two-phases: a silver-rich solid solution, α, containing 0.03% copper and a copper-rich solid solution, β, containing 0.975% copper.

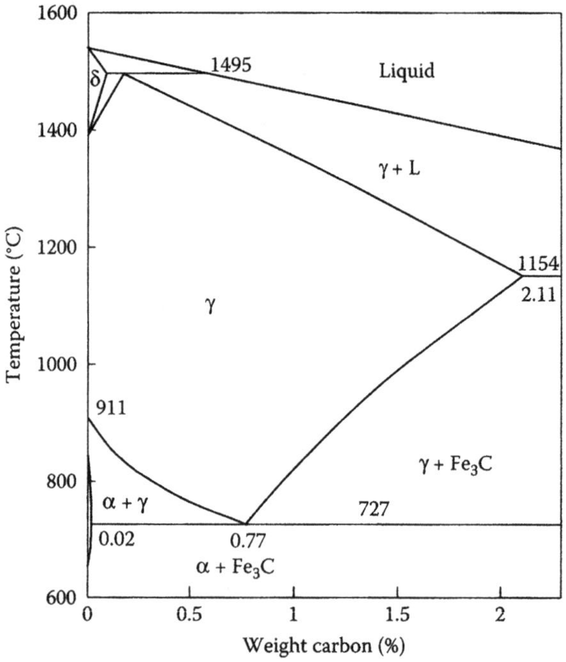

Fig. 2.8 The iron-carbon phase diagram. Source: Ref 2.3

In one gram of alloy, there is 0.8 gram of copper, divided between the two phases. If we designate the fraction of the microstructure, that is α by f_α, the amount of copper in the α-phase is $0.03f_\alpha$. Likewise there will be $0.975(1\text{-}f_\alpha)$ gram of copper in $(1\text{-}f_\alpha)$ grams of β-phase. A mass balance gives

$0.8 = 0.03f_\alpha + 0.975(1 - f_\alpha)$ or

$f_\alpha = (0.8 - 0.03)/(0.975 - 0.03) = 81.5\%$.

This procedure can be generalized as the lever law

$$f_\alpha = (c_{av} - c_\alpha)/(c_\beta - c_\alpha), \qquad \text{(Eq 1)}$$

where c_α and c_β are the compositions of the α and β-phases and c_{av} is the overall composition of the alloy.

Freezing

Almost all solid metals and alloys are produced from liquids by freezing (vapor deposition and electrodeposition are exceptions.) Often the

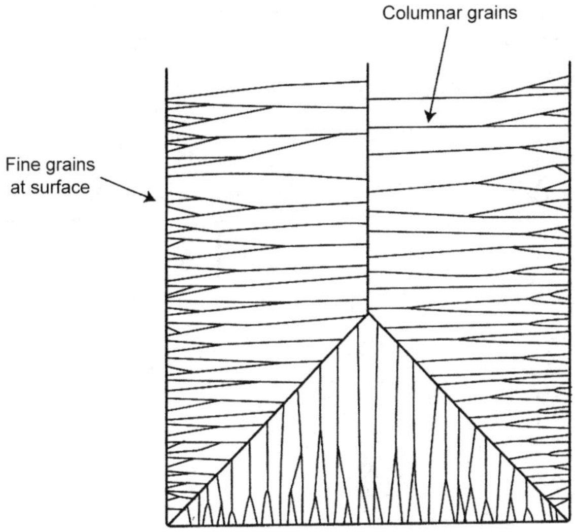

Fig. 2.9 Typical structure of a cast ingot with many small grains at the surface and columnar grains extending into the interior. Source: Ref 2.1

solidified metal is used in the form of a casting with little or no additional shaping. On the other hand, most metals are produced in the form of wrought products. These are solidified as ingots or continuously cast billets before being shaped into plate, sheet, rod, etc. For both castings and wrought products, what happens during freezing has a major influence on the microstructure and properties of the final product. Of particular importance are porosity and segregation of alloying elements.

The liquid state is intermediate between a gas and a crystal. Liquids do not have the high degree of order of crystals or the complete disorder of gases. For most materials, freezing causes a volume decrease of 1 to 6%. However, there are a few materials (water, silicon, germanium, bismuth, gallium, etc.) that have packing in the solid state that is not dense and that actually expands when they solidify.

Freezing starts at the outside of a casting, where the temperature is lowest. Many crystals may start to form on the outside surface, but the ones more favorably oriented for growth tend to dominate and cut off less favorably oriented crystals. This results in a columnar structure in the interior with finer grains at the surface, as shown in Fig. 2.9.

Segregation during Freezing

Solutes become segregated during the freezing of an alloy. Figure 2.10 shows that the first solid to form will be purer than the overall composition. As the system is cooled, newly formed solid will contain ever-increasing amounts of B. An unrealistically slow cooling rate (years or centuries per

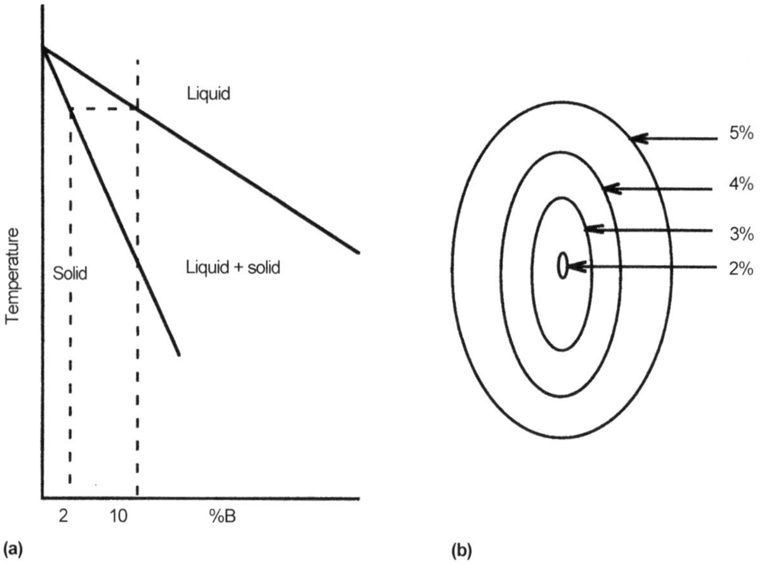

Fig. 2.10 Schematic drawing showing a binary phase diagram (a) and contours of concentration in a region that has frozen (b). Source: Ref 2.1

degree) would be required to allow enough time for diffusion to eliminate these concentration gradients.

This enrichment of the liquid near the solid-liquid interface causes the formation of a boundary layer in which liquid near the solid-liquid interface contains more solute than the liquid further away from the boundary (Fig. 2.11a). In the liquid there is a variation of the freezing temperature (liquidus temperature) just ahead of the interface, as shown in Fig. 2.11(b). These temperatures are determined by comparing the local composition with the phase diagram (Fig. 2.11c). The liquidus temperature and the actual temperature are plotted in Fig. 2.11 (b). At the interface, the actual temperature must equal the liquidus temperature according to the concept of local equilibrium. Just ahead of the interface, the actual temperature is lower than the liquidus temperature for the local composition. Such a condition is called *constitutional supercooling* because it results from compositional (constitutional) variations. This situation is not stable. If any area of the liquid interface happens to extend slightly ahead of the other areas, it will freeze faster and grow rapidly into the undercooled liquid, and dendrites will form with arms reaching out into the liquid.

Figure 2.12 shows two-dimensional dendrites forming in a polymer solution.

With dendrite formation, segregation occurs between the centers of the dendrite arms and the region between the dendrites. The liquid-solid shrinkage will cause inter-dendritic porosity rather than large cavities.

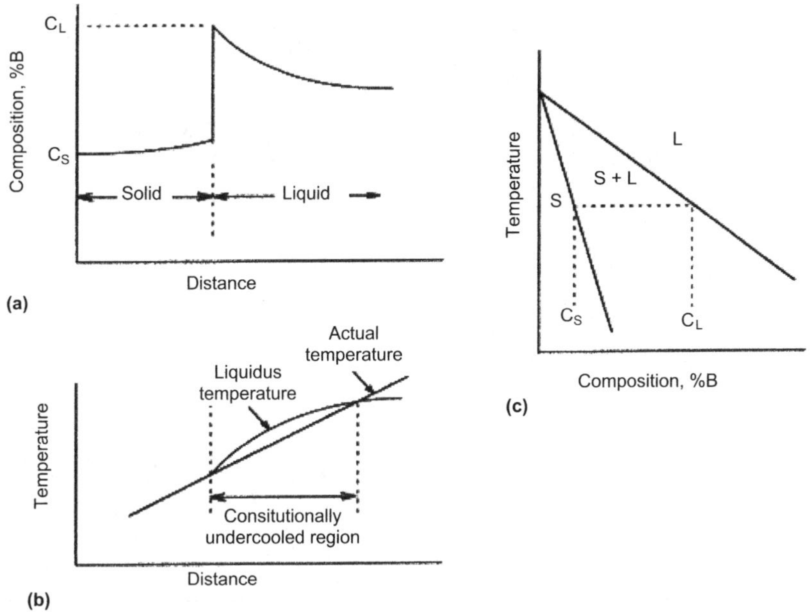

Fig. 2.11 Constitutional supercooling resulting from boundary layer formation. Source: Ref 2.1

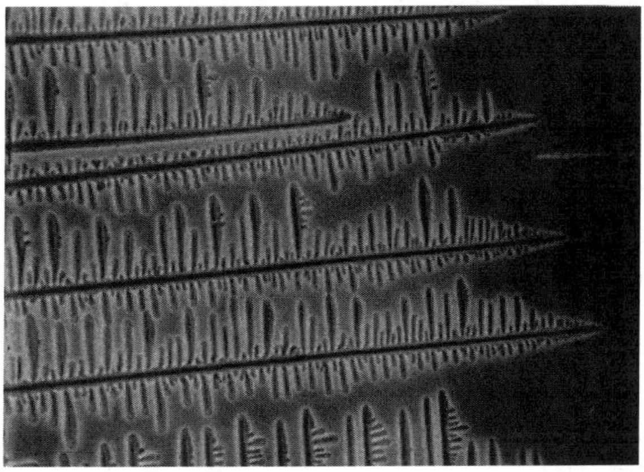

Fig. 2.12 Two-dimensional dendrites formed during the freezing of a polymer solution. Source: Ref 2.4

Metal Glasses

Materials that would form complex structures or several phases if they crystallized may not crystallize on freezing but instead may freeze as glasses. This is particularly true of polymers and complex silicates. There also are some metal compositions which freeze to glass structures if cooled fast enough. Most of the metal early compositions to form glasses included appreciable amounts of small atoms including phosphorus, silicon, beryllium, and germanium. Cooling rates of 10^3K/s to 10^5K/s were necessary to prevent crystallization. This limited glass formation in alloys to thin ribbons or wires. More recently magnesium-base, iron-base and zirconium-titanium-base alloys have been developed that do not require such rapid cooling. The first such commercial alloy available in bulk form is Vitreloy 1, which contains 41.2% Zr, 13.8% Ti, 2.5% Cu, 10% Ni, 22.5% Be. Because the critical cooling rate is about 1K/s, glassy parts can be made with dimensions of several centimeters. Metallic glasses can store a great deal of energy elastically. This accounts for their use in golf club heads as shown in Fig. 2.13.

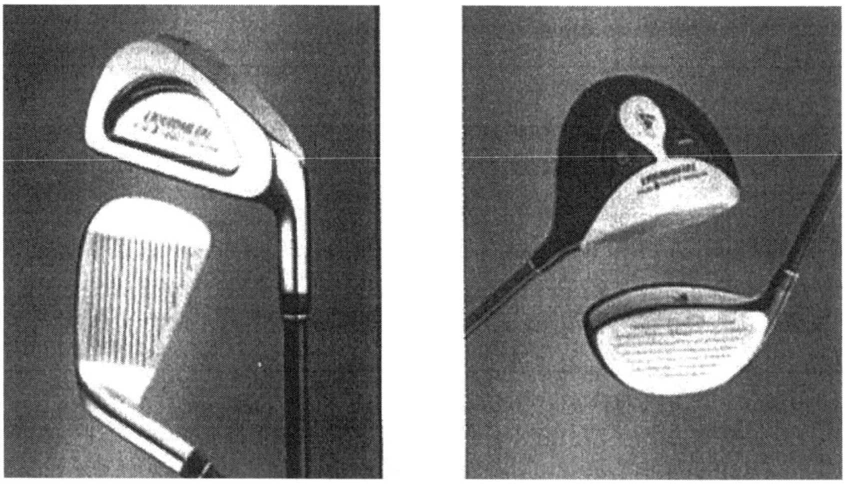

Fig. 2.13 Golf clubs made from metallic glass. Courtesy of: Otis Buchanan, Liquidmetal Technology, Lake Forrest, CA

NOTE OF INTEREST

Josiah Willard Gibbs (1839–1903) was an American theoretical chemist and physicist. He spent his entire professional career at Yale University, which awarded him the first American PhD in engineering in 1863. His paper, "On the Equilibrium of Heterogeneous Substances," published in 1876, formed the basis for the development of phase diagrams. As one of the greatest American scientists of the nineteenth century, his work laid much of the theoretical foundation for chemical thermodynamics. He invented vector analysis. Although his work was little known in America during his life, it was widely recognized abroad. His work inspired James Clerk Maxwell, a Scottish mathematical physicist and Gibbs' most enthusiastic supporter, to personally make a plaster cast illustrating Gibbs' construction that Yale University still owns.

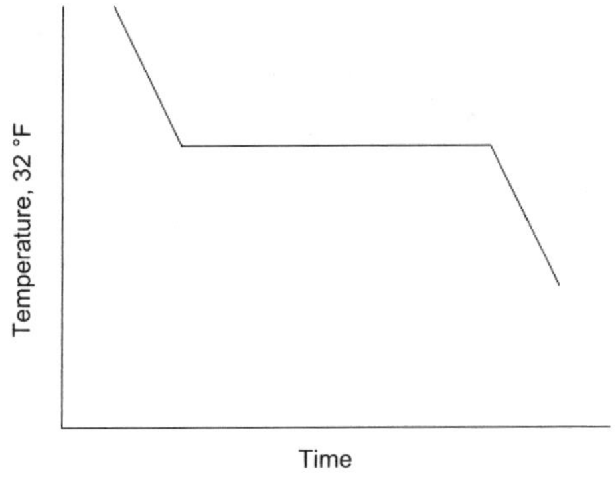

Fig. 2.14 Water is turning to ice during the horizontal hold

Exercises

1. **Density of Different Phases**

   Pour melted wax into a cup and allow it to freeze. Note the shape of the cavity that forms because solid wax is denser than liquid wax and therefore occupies less volume.

   One should see a cavity that is larger at the top than at the bottom.

2. **Temperature versus Time during Freezing of Water**

   Put a thermometer into a half pint of water and put it into a freezer. Record the temperature at intervals of five minutes and plot the temperature as a function of time. What is happening during the horizontal hold?

   A plot of temperature versus time should look like the diagram in Fig. 2.14.

REFERENCES

2.1 W.F. Hosford, *Physical Metallurgy,* 2nd ed., CRC Press, 2010
2.2 D.S. Clark and W.R. Varney, *Physical Metallurgy for Engineers* 2nd ed., Van Nostrand Reinholt, 1962
2.3 *Metals Handbook,* Vol. 8, 8th ed., ASM International, 1973
2.4 K.A. Jackson, *Solidification,* ASM International, 1971

CHAPTER 3

Mechanical Behavior

MECHANICAL BEHAVIOR of a material is how it deforms or breaks when it is stressed. When a force is applied to a solid material, there are three possible responses. The material may deform *elastically* which means it will regain its original shape when the force is removed. Springs and rubber bands deform elastically. If the force is high enough, the material will not regain its original shape and the result is *permanent* or *plastic* deformation. Dents in fenders of cars are examples of plastic deformation. A still greater force may cause the material to fracture.

Stress and Strain

Stress is defined as the intensity of force, F, per area. True stress, σ, is defined in terms of the current area, A,

$$\sigma = F/A.$$

Nominal or engineering stress, S, in a tension or a compression test is defined in terms of the original area, A_o,

$$S = F/A_o.$$

Strain describes the amount of deformation a material has undergone. Engineering or nominal strain, e, is defined simply as

$$e = \Delta L/L_o.$$

Elasticity

If a low stress is applied to a material, it will deform elastically and when the stress is removed the material will return to its original shape. Although stretching of a rubber band is a familiar example, most materials can undergo much less elastic deformation than rubber. The maximum elastic strain in crystalline materials usually is less than 0.2 %. For most materials (not rubber) it is safe to assume that the strain is proportional to the stress. The amount of elastic deformation is proportional to the stress, and can be described by

$$e = S/E,$$

where E is called the elastic modulus or Young's modulus.

When a material is stretched elastically, the bonds between neighboring atom, ions, or molecules are stretched. Bonding is strong in materials with high melting points. This strong bonding also is reflected in the elastic stiffness of a material; therefore, materials with high melting points tend to have high elastic moduli, as shown in Fig. 3.1.

Table 3.1 lists values of Young's moduli for several materials.

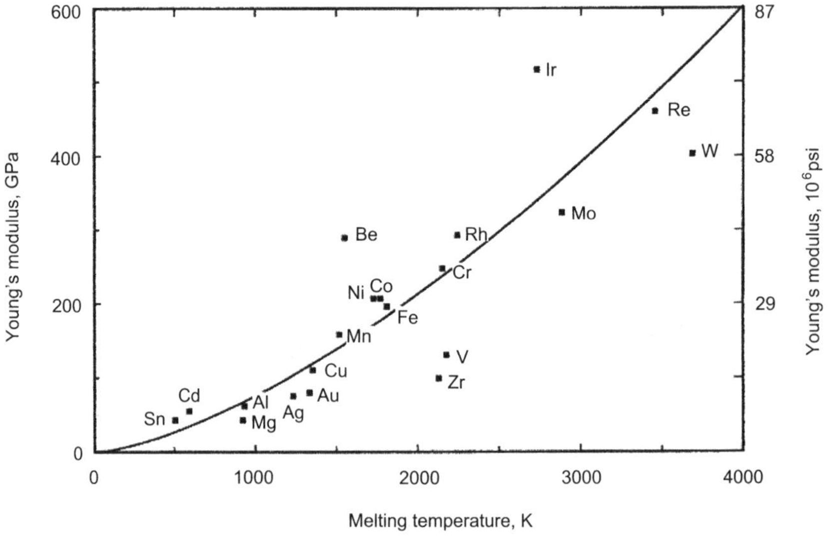

Fig. 3.1 Correlation of elastic moduli with melting points. Metals having a high melting point generally have a high elastic modulus. Source: Ref 3.1.

Table 3.1 Young's moduli at 20 °C (70 °F)

Material	Young's modulus GPa (10^6 psi)
Aluminum	70 (10)
Iron and steel	205 (29.5)
Copper	110 (16)
Magnesium	45 (6.5)
Titanium	116 (19)
Glass	70 (10)
Polystyrene	2.8 (0.5)
Nylon	2.8 (0.5)
LD polyethylene	0.1 – 0.35 (0.05 – 0.1)
Urea formaldehyde	10.3 (1.5)

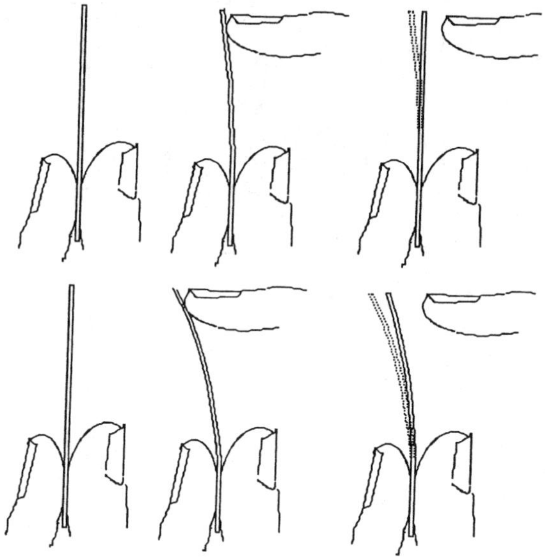

Fig. 3.2 Use of the fingers to sense the elastic and plastic response of a wire. With a low force (top) the deformation is entirely elastic and the bending disappears when the force is removed. With greater force (bottom) the elastic portion of the bending disappears when the force is removed but some plastic deformation remains. Source: Ref 3.2.

Plastic Deformation

If the amount of deformation exceeds a critical amount, atoms slide over one another and the material will not return to its original shape. In this case, the material is said to have *yielded* and deformed *plastically.* This is illustrated schematically in Fig. 3.2. Plastic behavior often is characterized by the engineering stress-strain curve in tension. Figure 3.3 is an engineering stress-strain curve of a ductile material.

It is tempting to define an *elastic limit* as the stress at which the very first plastic deformation occurs, and a *proportional limit* as the stress at which stress is no longer proportional to strain. However, both of these quantities depend on how accurately the stress and strain are measured. Increased accuracy of measurement results in lower values for the elastic and the proportional limits. To avoid this problem, it is customary to define an *offset yield strength* that can be measured reproducibly. The offset yield strength is found by constructing a straight line parallel to the initial linear portion of the stress-strain curve but offset from it by a strain of 0.002 or 0.2% as shown in Fig. 3.4. The stress at the intersection of this line with the stress-strain curve is taken as the 0.2% offset yield strength. Loading a tensile specimen to this stress and then unloading will result in a plastic strain of 0.002.

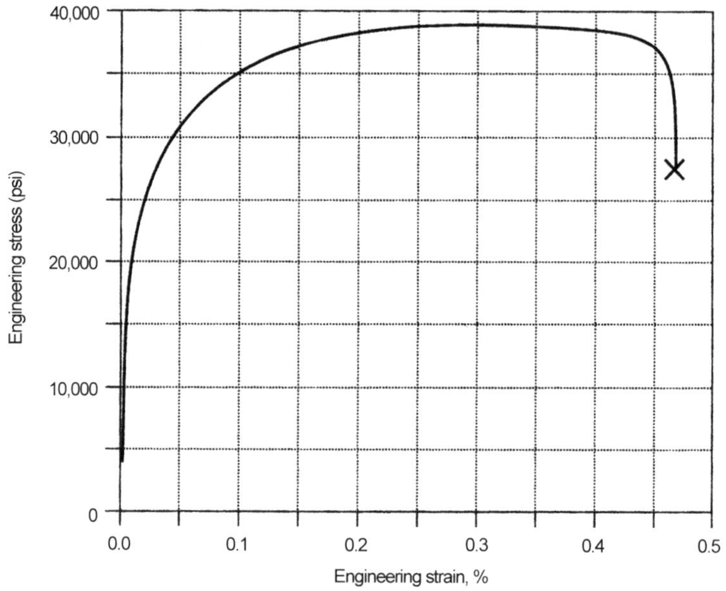

Fig. 3.3 Typical engineering stress-strain curve for a ductile material. Source: Ref 3.1.

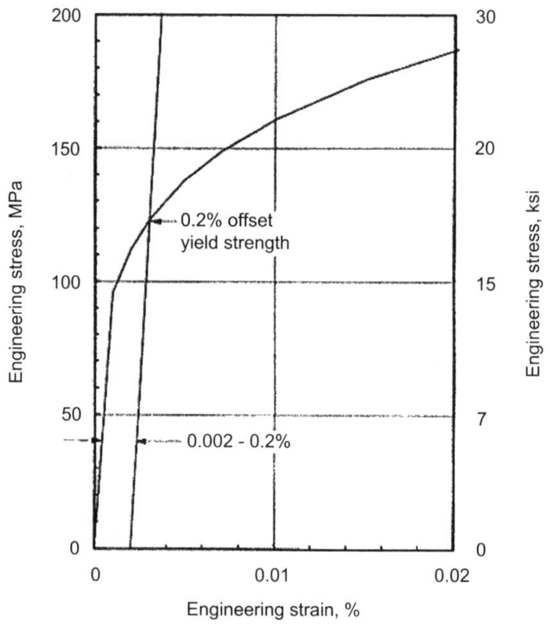

Fig. 3.4 The first portion of a stress-strain curve illustrating how the 0.2% offset yield strength is determined. Source: Ref 3.1.

For a few materials, an initial drop of stress occurs after the first yielding. Examples include low carbon steel (Fig. 3.5a) and linear polymers (Fig. 3.5b). For these materials, an *upper yield strength* is defined as the stress at the first maximum, and a *lower yield strength* is defined as the level of the plateau following initial yielding.

The highest engineering stress is called the *tensile strength* or *ultimate strength*. If a material is ductile, as in Fig. 3.6, its tensile strength is reached before it breaks. In this case its tensile strength corresponds to the point at which the deformation begins to localize to form a neck. For less ductile materials, fracture occurs before necking or even before yielding; therefore, the tensile strength is the fracture strength (Fig. 3.7).

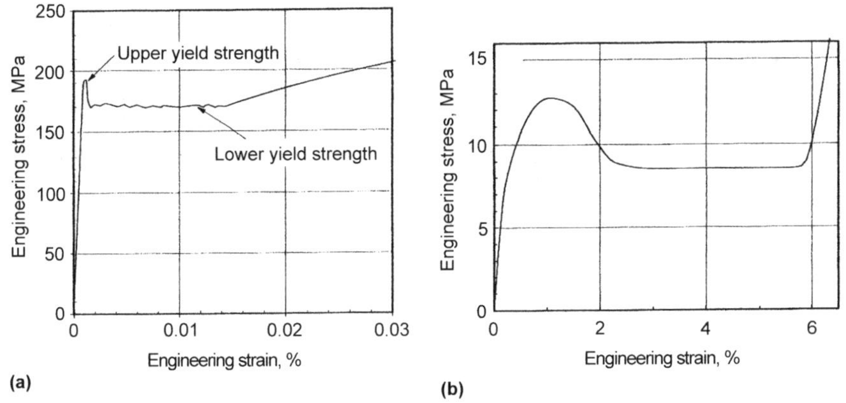

Fig. 3.5 Inhomogeneous yielding of low carbon steel (a) and a linear polymer (b). After the initial stress maximum, the deformation in both materials occurs within a narrow band that propagates the length of the gage section before the stress rises again.

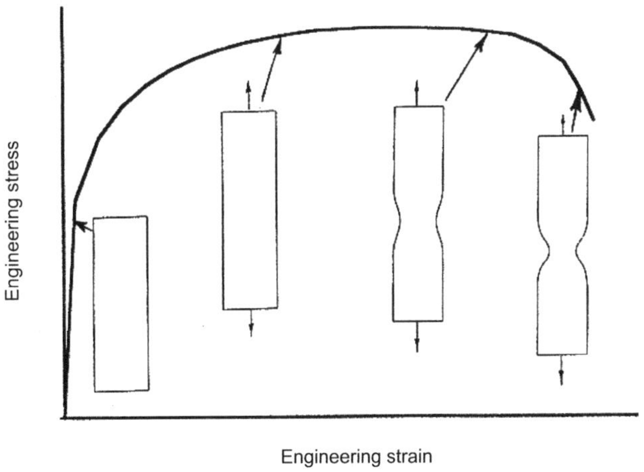

Fig. 3.6 Necking starts when a maximum engineering stress is reached. Source: Ref 3.1.

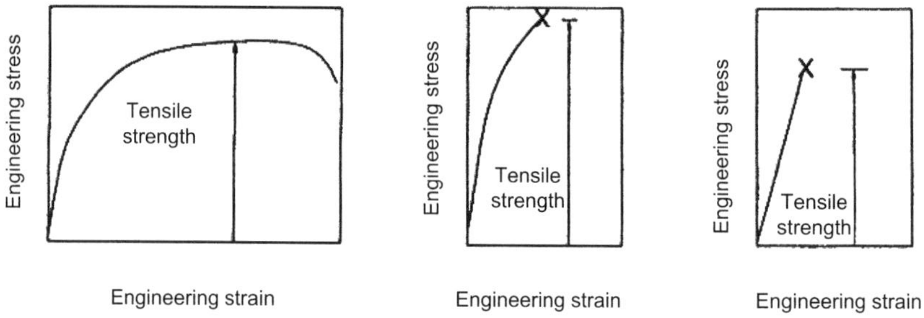

Fig. 3.7 The tensile strength is the maximum engineering stress in a tension test, regardless of whether the specimen necks, fractures before necking, or fractures before yielding. Source: Ref 3.1.

Ductility

The ductility of a material describes the amount of deformation before fracture in a tension test. There are two common measures of ductility. One measure is the *percent elongation* before fracture,

$$\%El = (L_f - L_o)/L_o \times 100\%.$$

The other is the *percent reduction of area,*

$$\%RA = (A_o - A_f)/A_o \times 100\%.$$

The percent reduction of area at fracture is a better measure of ductility because the percent elongation involves both the uniform elongation before necking and the elongation during necking so that the percent elongation during necking depends on the ratio of the specimen diameter to its length. Tensile specimens with large length-to-diameter ratios have lower % elongations than those with low length-to-diameter ratios.

It should be realized that the amount of deformation that a material can withstand depends on the form of loading. Under compression, as in rolling, the ductility is far higher than that measured in a tension test.

Hardness

The hardness test is a simple way of characterizing a material's resistance to plastic deformation. In a hardness test, an indenter is pressed into the surface of a material under a fixed force and the extent of indentation is measured. In the Brinell, Vickers, and Knoop hardness tests, the dimensions of the indentation are measured after the load is removed; the hardness equals the load divided by the area of the indentation. With the several Rockwell tests, the depth of the indentation is automatically recorded as the hardness number. Materials with high hardnesses have high yield strengths.

Creep

At high temperatures, materials under stress may deform slowly. Such time-dependent deformation is called *creep*. Figure 3.8 shows that the rate of creep increases with increasing stress and temperature.

Fatigue

Failures may occur under repeated applications of stress at a level which would not cause failure if the stress were applied only once. This phenomenon is called *fatigue*. The number of cycles before failure increases as the stress is lowered, as shown in Fig. 3.9. For steels there is a level of stress

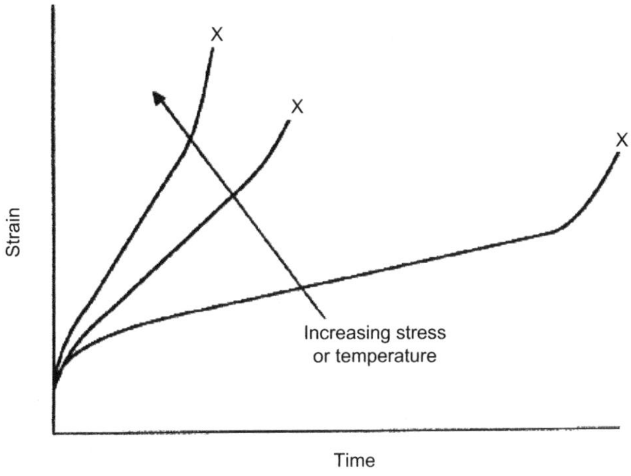

Fig. 3.8 Increased stress and increased temperature cause faster creep

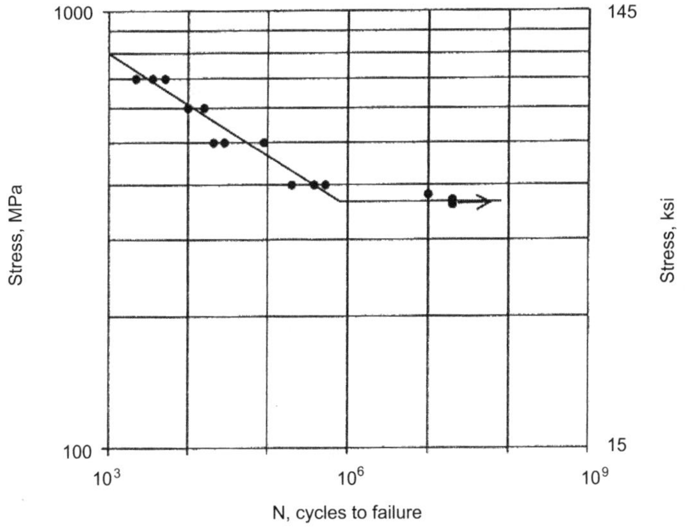

Fig.3.9 Fatigue behavior of steels

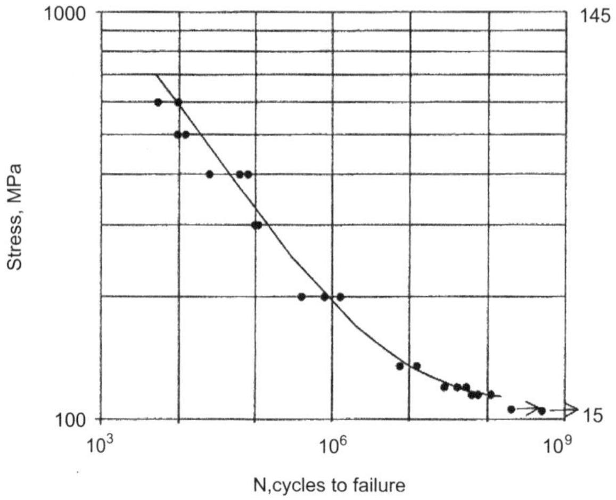

Fig. 3.10 Fatigue behavior of aluminum

below which fatigue failure will not occur even after an infinite number of cycles. This is called the *endurance limit* or the *fatigue strength*. For nonferrous metals there is no such fatigue limit, as shown in Fig. 3.10.

Fracture

The mode in which a fracture occurs may be classified as ductile or brittle, depending on how much deformation precedes the fracture. Fractures may occur along grain boundaries or, more commonly, through the grains. The tendency of a material to break in a brittle manner is increased by low temperatures, high rates of loading, and the presence of notches. These three factors tend to suppress plastic deformation; therefore, permitting higher stresses. Molasses candy is a good example of these effects. If one tries to break it by bending, it just deforms. However, if it is refrigerated, notched, and rapped sharply, it is likely to break before it deforms. These effects are illustrated in Fig. 3.11.

A material is regarded as being *tough* if it absorbs a large amount of energy in breaking. In a tension test, the energy per volume to cause failure is the area under the stress-strain curve and is the toughness in a tension test. However, the toughness under other forms of loading may be very different because toughness depends also on the degree to which deformation localizes. The total energy to cause failure depends on the deforming volume as well as on energy per volume.

Impact tests often are used to assess the toughness of materials. The most common of these is the *Charpy test*. A notched bar is broken by a swinging pendulum. The energy absorbed in the fracture is measured by recording how high the pendulum swings after the bar breaks. Figure 3.12 gives the details of the test geometry. The standard specimen has a

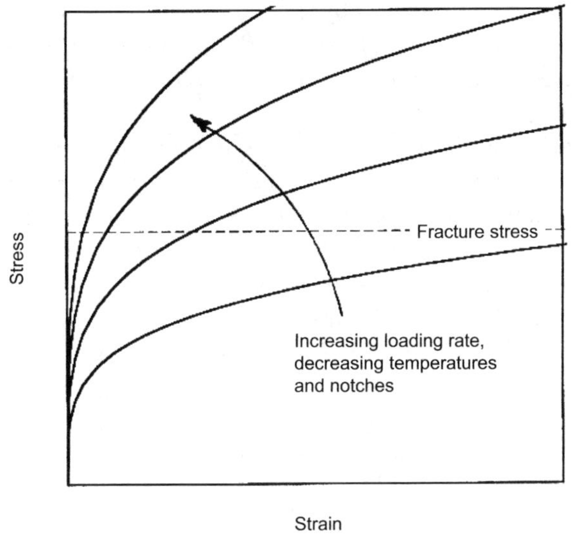

Fig. 3.11 Increased rates of loading, lower temperatures, and the presence of notches tend to make materials behave in a brittle manner because they raise the stresses to a level where fracture can occur

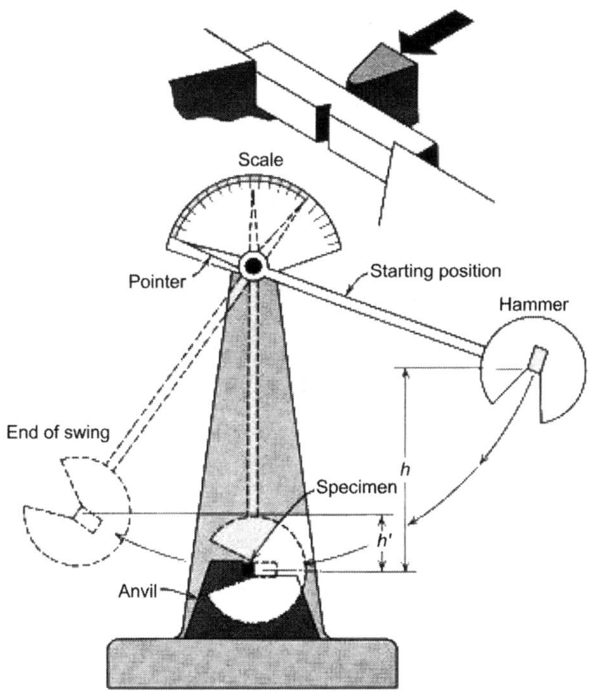

Fig. 3.12 A Charpy testing machine and test bar. A hammer on the pendulum breaks the bar. The height the pendulum swings after breaking the bar indicates the energy absorbed. Source: Ref 3.3.

cross section 10 by 10 mm (0.5 by 0.5 in.). There is a 2 mm (0.05 in.) deep V-notch with a radius of 0.25 mm (0.01 in.). The pendulum's mass and height are standardized. Sometimes bars with U- or keyhole notches are employed instead.

The toughness in a Charpy test can be measured easily over a range of temperatures. A specimen can be heated or cooled to the specified temperature, and then transferred to the Charpy machine and broken quickly enough so that its temperature change is negligible. For many materials there is a narrow temperature range, over which there is a large change of energy absorption and fracture appearance. It is common to define a *transition temperature* in this range. At temperatures below the transition temperature, the fracture is brittle and absorbs little energy in a Charpy test. Above the transition temperature the fracture is ductile and absorbs a large amount of energy. Figure 3.13 shows typical results for steel.

The transition temperature *does not* correspond to any structural change of the material. The ductile-brittle transition temperature depends greatly on the type of test being performed. Lower transition temperatures are measured with less severe notches. With slow bend tests and un-notched tensile tests, even lower ductile-brittle transitions are observed. In discussing the ductile-brittle transition temperature of a material, one should specify not only the type of test, but also the criterion used. Specifications for ship steels often require a certain *Charpy V-notch 15 ft-lb transition*

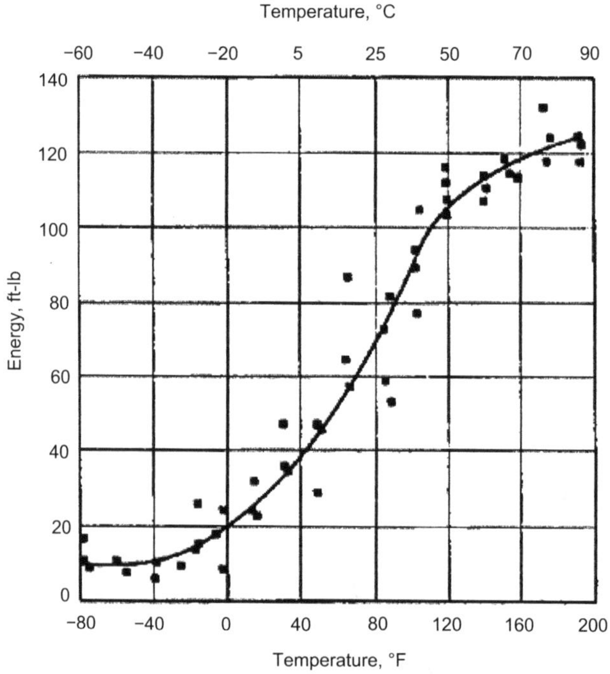

Fig. 3.13 Ductile-brittle transition in a Charpy V-notch specimen of a low-carbon, low-alloy, hot-rolled steel. Source: Ref 3.4.

temperature (the temperature at which the energy absorption in a V-notch Charpy test is 15 ft-lbs).

In general, bcc metals and many hcp metals exhibit a ductile-brittle transition, but it is significant that fcc metals do not. For fcc metals, changes of impact energy with temperature are small. Because austenitic stainless steels and copper are not embrittled by low temperatures, they frequently are used in equipment for applications that involve low temperatures.

In the period 1948 to 1951, there were many fractures of natural gas pipelines. Most occurred during testing, and most started at welding defects but propagated through sound metal. One of the longest cracks was 975 m (3200 ft) long. Once started, such cracks run at speeds greater than the velocity of sound in the pressurized gas. Therefore, there is no release of the gas pressure to reduce the stress at the tip of the crack. Figure 3.14 is a photograph of one of the cracked lines.

During World War II, there was a rapid increase of shipbuilding. The production of ships by welding together steel plates (in contrast to the earlier process of joining them by riveting) became common during World War II. As a result, a large number of ships, particularly Liberty ships and T-2 tankers, failed at sea. More ships sunk as the result of brittle fractures than by German U-boat attacks. Recovery of some ships and half-ships allowed the cause of the failures to be investigated. It was found that there were three main factors: poor welds, bad ship design (cracks often started at sharp cornered hatchways that created stress concentrations), and high

Fig. 3.14 A natural gas pipeline that failed during field testing. Source: Ref 3.5.

> **NOTE OF INTEREST**
>
> The Great Boston Molasses Flood occurred on January 15, 1919 in Boston. A large molasses storage tank suddenly burst and a wave of molasses ran through the streets at an estimated 35 mph, killing 21 and injuring 150. Residents claim that on hot summer days the area still smells of molasses. This was the first brittle fracture failure of a large structure. Other failures of large engineering structures include fractures in pipelines and ships.

Fig. 3.15 A ship that fractured while still in port. Source: Ref 3.6.

transition temperatures of the steels. Figure 3.15 is a photograph of a ship that failed in harbor.

Exercises

1. **Stress-Strain Curve Differences**

   Take a piece of iron florists wire and bend it by holding the ends. All of the bend will occur in a short region.

   Now repeat the experiment with a piece of copper wire. You should see that the bend is distributed over the whole wire.

   Note: You may have to anneal each wire before the experiment by putting it in a 230 °C (450 °F) oven for a few minutes.

The different behavior is the result of the differences in the stress-strain curves of the two metals. With the iron wire, initial yielding in one region makes that region softer than the other regions, so the deformation will localize there. The copper wire strain hardens, causing regions that have yielded to be harder to deform than other regions, so the deformation will spread over the whole wire.

2. **Metal Fatigue**

   Take a paper clip and bend it. Then unbend it. Note that unbending it requires more force than bending it. Also note that if the paper clip is bent back and forth repeatedly, it eventually will break.

   This is an example of fatigue.

3. **Cardboard Fiber Alignment**

   Cut a piece of cardboard into a square, orient it so it is parallel to the floor, and support it at two opposite edges. Place a weight in the center and note the deflection. Remove the weight, rotate the cardboard square 90° so it is supported at the other two edges, replace the weight, and note the deflection.

   The difference in stiffness is caused by the alignment of fibers during the production of the cardboard.

4. **Paper Fiber Alignment**

   Tear a newspaper vertically and then horizontally.

   The fact that one tear is almost straight and the other wanders is caused by the alignment of fibers during the production of the paper.

5. **Effects on Ductility**

   Take a piece of molasses candy and try to break it slowly. Then try to break it by rapping it suddenly on a hard surface. Then cool it in a freezer and try to break it. Finally, scratch a notch into it and try to break it. What effects do low temperatures, high rates of loading, and notches have on the ductility?

   Applying a high rate of loading (rapping suddenly), cooling, or notching lowers the ductility of the molasses, causing it to break.

REFERENCES

3.1 W.F. Hosford, *Mechanical Behavior of Materials,* 2nd ed. Cambridge University Press, 2010

3.2 W.F. Hosford in *Tensile Testing,* ASM International, 1992
3.3 H.W. Hayden, W.G. Moffatt and J. Wulff, *Structure and Properties of Materials, Vol. III Mechanical Behavior,* Wiley, 1965.
3.4 R.W. Vanderbeck and M. Gensamer, *Welding J. Res. Suppl.* p. 37s–48s, January 1950
3.5 E. Parker, *Brittle Fracture of Engineering Structures,* Wiley, 1957
3.6 C.F. Tipper, *The Brittle Fracture Story,* Cambridge University Press, 1963

CHAPTER 4

Electrical Behavior

Conduction

Electrical conduction through materials occurs by the movement of charge carriers such as electrons, missing electrons, or ions. The conductivity, σ, of a material is the product of the number of charge carriers per volume, n, the charge on the carriers, q, and their mobility, μ,

$$\sigma = n\mu q$$

Mobility is the ratio of drift velocity to the electric field.

Metallic Conduction

Metals are excellent conductors. Table 4.1 lists the conductivities of some metals. Temperature and impurities affect the conductivities of metals. Their effects can be understood in terms of how they influence mobility. Valence electrons form an electron gas that is free to move throughout a metal. They are in constant motion. If there is no applied field, their movement is random. However, an electric field accelerates the electrons in the direction of the field, causing a net drift velocity. Any irregularity in the lattice will cause a random change in the path. The drift velocity increases

Table 4.1 Electrical conductivities at 20 °C (70 °F)

Metal	Conductivity $(n\Omega m)^{-1}$
Aluminum	37.7
Copper	59.77
Silver	68.0
Nickel	14.6
Iron	10.5
Magnesium	22.47
Zinc	16.9
70/30 brass	16.7
Cupronickel (20% Ni)	3.9
Stainless steel (18% Cr,8% Ni)	13.9

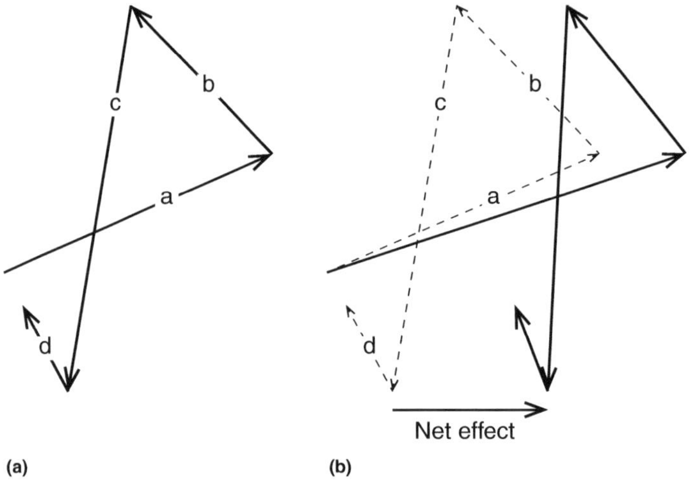

Fig. 4.1 Imposition of an electrical field on the random motion of electrons creates a drift velocity. Random motion (a) and the influence of field (b). Source: Ref 4.1.

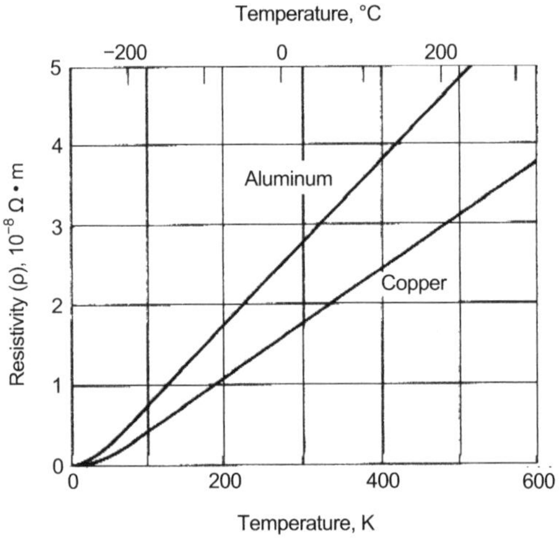

Fig. 4.2 The electrical resistivities of copper and aluminum vary linearly with temperatures above about 100 K. Source: Ref 4.2.

with the field and the mean free path between interactions with lattice irregularities. Therefore, the conductivity and mobility are determined by the mean free path. Figure 4.1 illustrates this concept.

With increased temperature, atoms vibrate more from their normal lattice positions, making the lattice less regular and decreasing the length of the mean free path. This decreases the drift velocity. Figure 4.2 shows that the resistivity, which is the reciprocal of conductivity, varies linearly with temperature, except at very low temperatures. Figure 4.3 shows the effects of various solutes on the resistivity of copper.

Cold working a metal creates lattice defects that raise the resistivity of the metal. Vacancies are the most important defects. The effect of each vacant lattice site acts the same way as a solute atom. Therefore, the change of resistivity caused by cold work, $\Delta\rho_{cw}$, is proportional to the number of vacancies. The effects of cold work, annealing, and temperature on resistivity are shown in Fig. 4.4.

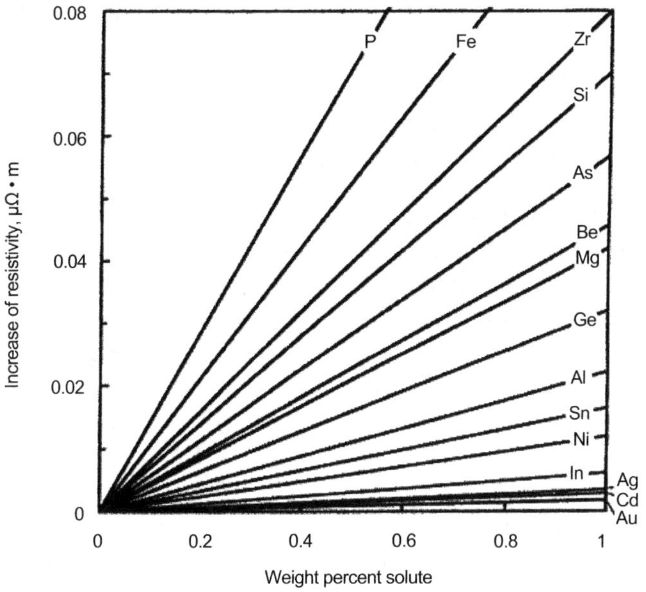

Fig. 4.3 Effects of various solutes on the conductivity of copper. Source: Ref 4.3.

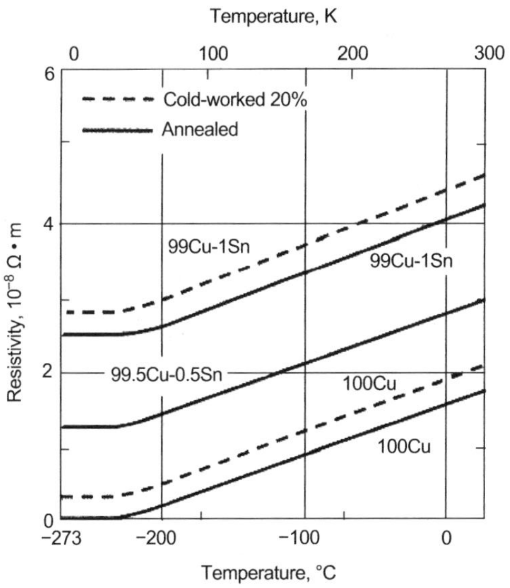

Fig. 4.4 Effects of annealing and cold work on the resistivity of copper. Source: Ref 4.4

Whereas increased temperature and impurities lower the conductivity of metals, they have the opposite effect in insulators and semiconductors.

Ionic Conduction

In ionic solids, current is carried by the movement of ions under the influence of an electric field, just as it can be in liquid solutions. The difference is that the rate of movement of ions in solids is much slower than in liquids. Increases of both temperature and impurity content increase conduction in ionic crystals.

Energy Bands

The difference between metals, semiconductors, and insulators can be understood in terms of the energy levels of the outer electrons. Electrons in isolated atoms can exist in only discrete energy levels, with no more than two electrons occupying any level. However, as a large number of atoms are brought together to form a crystal, the electric fields of the various atoms interact, causing permissible energy levels to split into bands. The number of electrons in each energy band can be no more than twice the number of atoms.

Because conduction requires the energy of electrons to increase, for any conduction, there must be available energy levels. With metals, either the valence band overlaps the conduction band or is only partially filled, thus there is a very large number of carriers. In semiconductors, the valence band is filled, but there is only a small energy gap between the valence and conduction bands. With insulators, the valence band is filled, and there is a large energy gap between the valence and conduction bands. Figure 4.5 illustrates these possibilities.

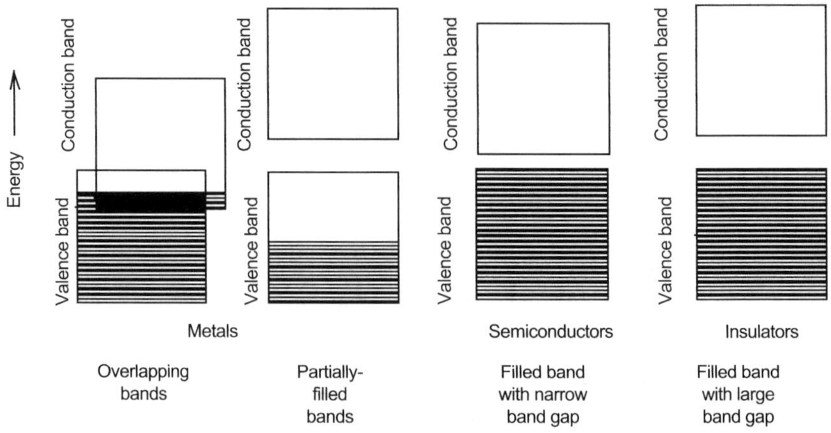

Fig. 4.5 The band structures of different classes of materials. Source: Ref 4.1.

Intrinsic Semiconduction

In a semiconductor like silicon, the energy gap is so small that thermal agitation will increase the energy of a small number, n, of electrons enough to promote them into the conduction band, as sketched in Fig. 4.6. The missing electron in the valence band is called an electron hole; it also can move through the lattice causing conduction. For a pure (*intrinsic*) semiconductor, the number of carriers increases with temperature.

Figure 4.7 illustrates this on an atomic level. For every electron promoted to the conduction band, a hole is generated in the valence band.

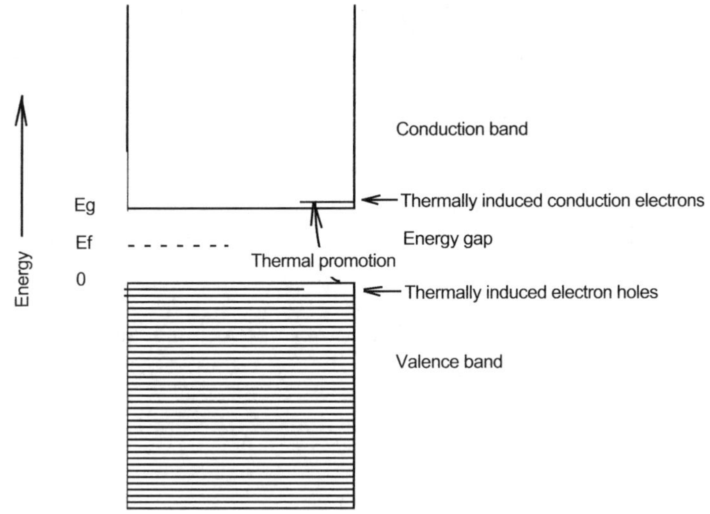

Fig. 4.6 Band structure of an intrinsic semiconductor. A few electrons are thermally promoted from the valence band to the conduction band. Source: Ref 4.1.

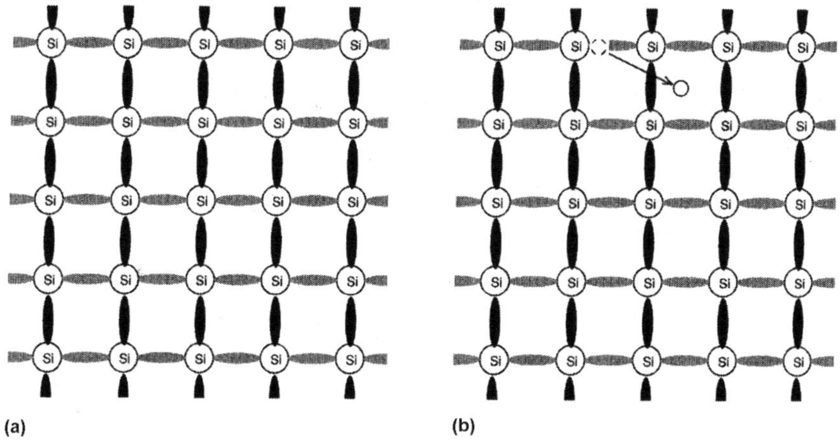

Fig. 4.7 (a) Each atom in silicon shares two electrons with each of four near neighbors. (b) If an electron is removed from bonding, a conduction electron and an electron hole are created. Source: Ref 4.1.

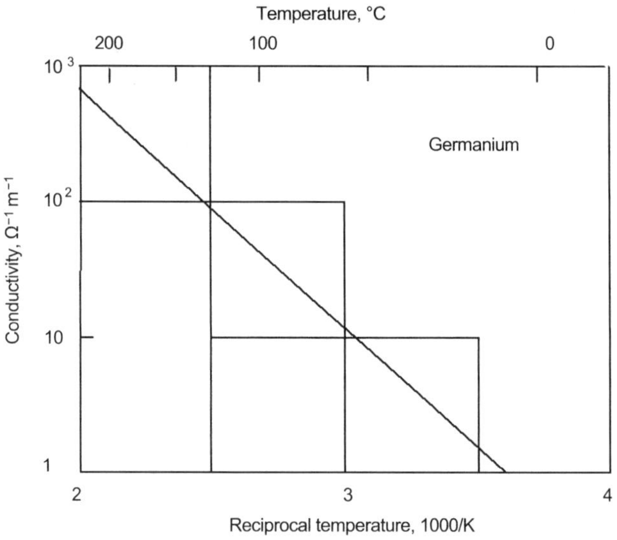

Fig. 4.8 The conductivity of pure germanium increases exponentially with temperature. Source: Ref 4.1.

Table 4.2 Energy gaps and mobilities in intrinsic semiconductors

		Mobility, m^2/V-s	
Semiconductor	Energy gap, eV	Electrons	Holes
Silicon	1.1	0.19	0.425
Germanium	0.7	0.36	0.23
GaAs	0.7	0.60	0.08
InAs	0.36	2.26	0.26
InP	1.3	0.47	0.015

The electrons in the conduction band are carriers of a negative charge, and the electron holes in the valence band act as carriers of a positive charge.

Figure 4.8 shows that the conductivity of germanium increases exponentially with temperature. The energy gaps and mobilities in several semiconductors are listed in Table 4.2.

Extrinsic Semiconduction

The addition of even very small amounts of impurities to a semiconductor greatly increases its conductivity. For example, impurities of group V elements (nitrogen, phosphorus, arsenic, antimony, bismuth) add an extra electron, which can act as a conductor. The extra electron creates a donor state with an energy level only slightly below the conduction band, as illustrated in Fig. 4.9, so very little thermal energy ($E_g - E_d$) is required to promote an electron from the donor state to the conduction band. At room temperature, almost all of the donor electrons are promoted. In this case, the conductivity primarily is by donor electrons; their number is approximately the number of group V atoms per volume, and the semiconductor is *n-type* (negative carriers). Figure 4.10 shows the temperature dependence in this case.

The effect of group III elements (aluminum, gallium, indium) is similar. It takes very little thermal energy to promote an electron from the valence band to the acceptor state. This creates a hole in the valence band (Fig. 4.11). At room temperature the acceptor state is almost filled so the number of electrons holes is approximately the number of group III atoms per volume and the semiconductor is *p-type* (positive carriers).

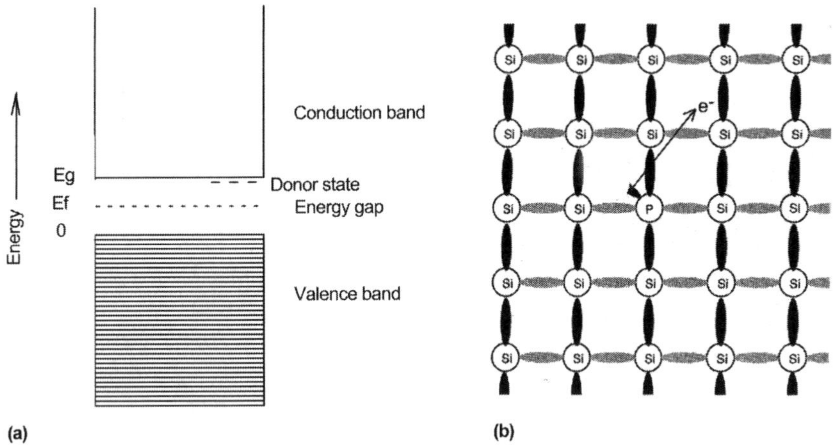

Fig. 4.9 Group V impurities create donors. (a) The energy level of donors is close to the conduction band. (b) The extra electron can easily be promoted to conduction. Source: Ref 4.1.

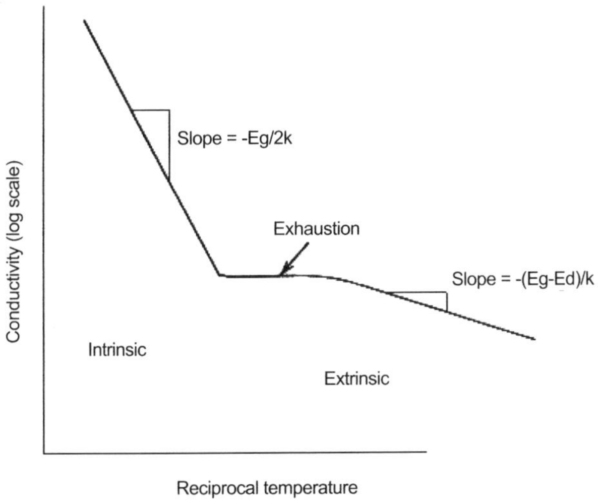

Fig. 4.10 The temperature dependence of conductivity in an extrinsic semiconductor. At very low temperatures, conductivity depends on thermal excitation of donor or acceptor levels. Near room temperature, there is saturation. At high temperatures, intrinsic conduction prevails. Source: Ref 4.1.

In the exhaustion range, the conductivity is approximately proportional to the impurity concentration.

III-V Compound Semiconductors

Compounds of group III and group V elements are semiconductors. Examples include aluminum antimonide, gallium phosphide, gallium arsenide, indium phosphide and indium antimonide. Zinc sulfide (group II–VI) also is a semiconductor. Either nonstoichiometric compositions or foreign impurities create *n*- or *p*-type semiconductors. Figure 4.12 illus-

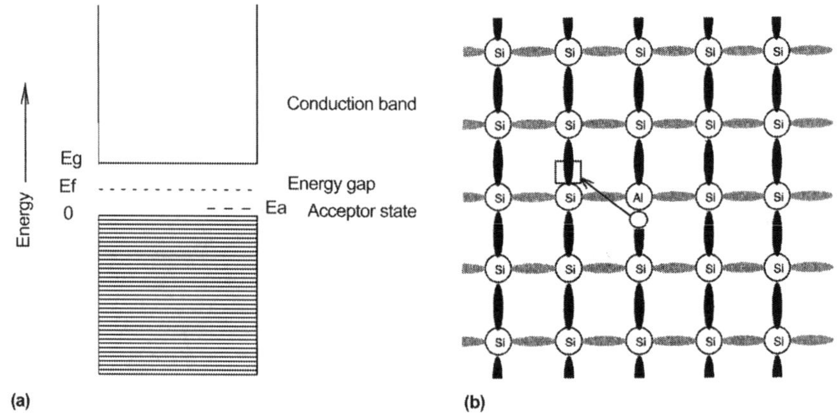

Fig. 4.11 Group III impurities create electron holes. (a)The acceptor energy level is slightly above the valence band. (b)The missing electron is easily promoted to be an electron hole. Source: Ref 4.1.

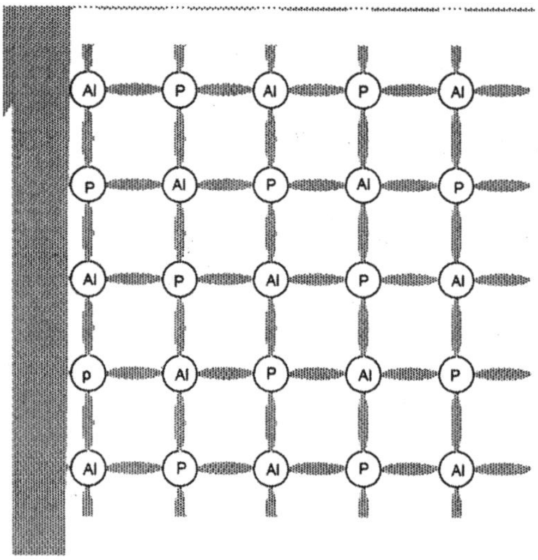

Fig. 4.12 Structure of a III-V semiconductor. Source: Ref 4.1.

trates schematically the structure of these compounds. Table 4.3 lists some of the properties of semiconductors.

p-n Rectifiers

A *rectifier* or *diode* allows current to pass in only one direction. A *p-n* junction can act as a rectifier. Figure 4.13 shows that if a forward bias is applied to the junction, holes in the *p*-material and electrons in the *n*-type material will flow to the junction where they will combine, allowing current to flow. More electrons and more holes are created where external

Table 4.3 Properties of some semiconductors

Material	Energy gap, E_g 10^{-18}J	Energy gap, E_g eV	Mobility, m^2/vs Electrons μ_n	Mobility, m^2/vs Holes μ_n	Intrinsic conductivity $(\Omega.m)^{-1}$ nm	Lattice parameter
Silicon	0.176	1.1	0.19	0.0425	5×10^{-16}	0.543
Germanium	0.112	0.7	0.36	0.23	2	0.566
AlSb	0.26	1.6	0.02			0.613
GaP	0.37	2.3	0.019	0.012		0.545
GaAs	0.22	1.4	0.88	0.04	10^{-6}	0.565
InP	0.21	1.3	0.47	0.015	500	0.587
InAs	0.058	0.36	2.26	0.026	10^{-4}	0.604
InSb	0.029	0.18	8.2	0.17		0.648
ZnS	0.59	3.7	0.014	0.0005		

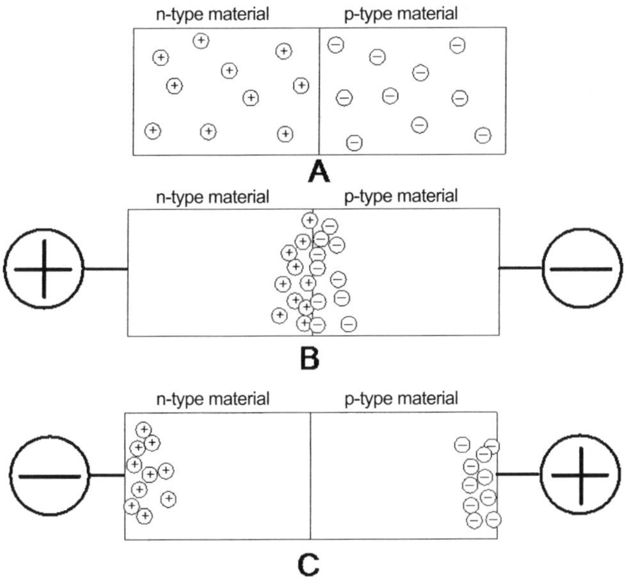

Fig. 4.13 A *p-n* junction rectifier has both *p*- and *n*-type regions (A). With a forward bias (B) both electrons and holes move toward the junction, allowing current to pass. With a negative bias (C) the region near the junction becomes depleted in carriers so no current can flow. Source: Ref 4.1.

leads are connected. With a reverse bias, electrons and holes move away from the junction creating a zone depleted of carriers.

Light Emitting Diodes

If a forward bias is applied across a p-n junction, electrons and holes annihilate each other. The energy of this recombination is converted into light. The wavelengths of the light of such light emitting diodes (LEDs) depend on the energy gaps. Table 4.4 lists the wavelengths of several compounds. These can be adjusted by varying the ratio of the elements.

Transistors

Junction *transistors* are either of the *n-p-n* type or the *p-n-p* type. In a *p-n-p* transistor (Fig. 4.14), the *n*-type material forms a thin base. Because the base is so thin, holes in the emitter can cross into the collector. A small input voltage signal controls the number of such holes as well as the collector current. The voltage drop across the large resistor in the collector circuit amplifies the smaller input voltage signal.

The other type of transistor is the metal oxide semiconductor field effect transistor (MOSFET) sketched in Fig. 4.15. The gate, made from polycrys-

Table 4.4 Common Light Emitting Diodes

Compound	Wavelength, nm	Color
GaP	565	Green
GaAsP	590	Yellow
GaAsP	632	Orange
GaAsP	649	Red

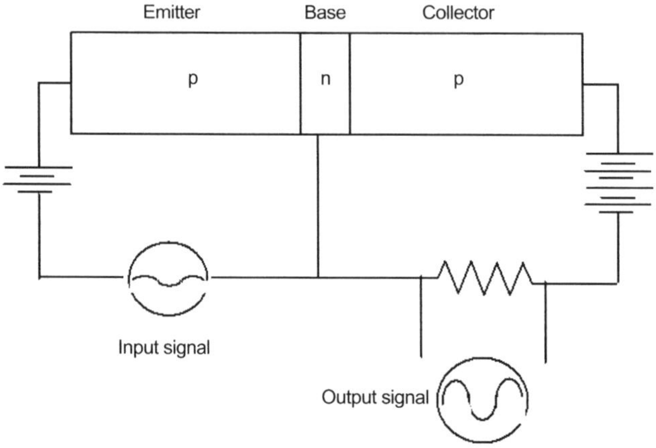

Fig. 4.14 A *p-n-p* transistor. A small input signal imposed on the emitter causes holes to penetrate the base and allows current in the collector. Because the voltage across the collector is high, the output signal is at high voltage. Source: Ref 4.1.

talline silicon, is separated from the channel by a layer of SiO_2. A small signal voltage between the source and gate creates an electric field that penetrates the channel and controls its conductivity. Thus modulating the field between the source and gate will modulate the conduction between the source and drain.

Polar Compounds

Some crystals are polarized. This happens if their centers of positive and negative charge do not coincide. One example is quartz. Another is barium titanate. Above 120 °C (250 °F), $BaTiO_3$ is cubic with the Ti^{4+} ion in the center of a cubic unit cell with O^{2-} ions in face-centering positions and Ba^{2+} ions at the corners. However, at lower temperatures, the cell becomes tetragonal. The Ti^{4+} ion shifts in one direction and the O^{2-} ions in the opposite direction relative to the Ba^{2+} ions at the corners. These displacements of the centers of positive and negative charge set up a permanent dipole. Electrons and the negative ions are attracted to the positive electrode. There is no voltage in the absence of an external field (Fig. 4.16) because the accumulated charge just balances the dipole.

Piezoelectric Behavior

Piezoelectricity of some polar crystals generates a voltage in response to applied stress and conversely to undergo a reversible strain in response to a voltage. The word piezoelectricity comes from the Greek *piezein*, which means to squeeze or press. The effect finds useful applications such as the production and detection of sound, generation of high voltages, electronic frequency generation, sonar, and ultra fine focusing of optical assemblies.

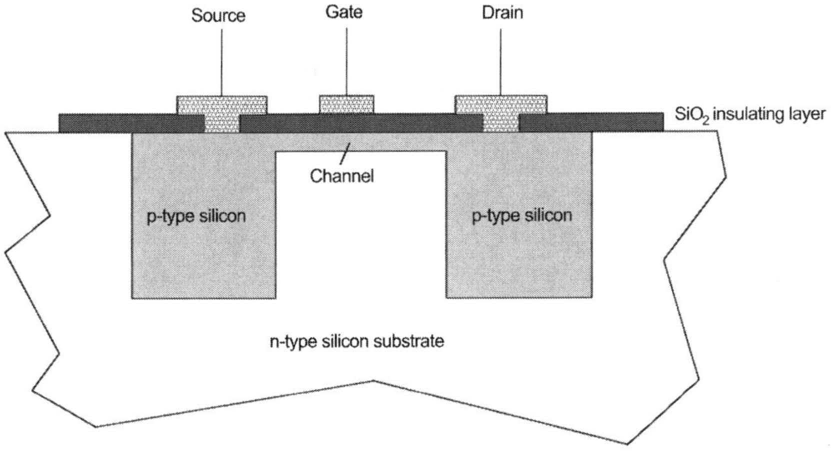

Fig. 4.15 A MOSFET transistor. Source: Ref 4.1.

An applied external field distorts the material, as indicated in Fig. 4.17, as the internal negative charge is attracted to the additional positive voltage. Conversely, an external force that strains the crystal produces a voltage.

Piezoelectric materials are useful in applications where mechanical vibrations need to be developed by vibrating fields such as sending units of sonar, audio speaker systems, and analog watches; and for high frequency sound generators. They also are used where mechanical vibrations are used to produce electrical signals as in the receiving units of sonar, microphones, and frequency standards. Above 120 °C (250 °F) (its Curie temperature), barium titanate has a cubic crystal structure that is not piezoelectric. Lead zirconate titanate (PZT), [$Pb(Zr,Ti)O_3$] is piezoelectric up to 480 °C (900 °F). Lead metaniobate has a Curie temperature of 570 °C (1060 °F). The most widely used piezoelectric material is PZT.

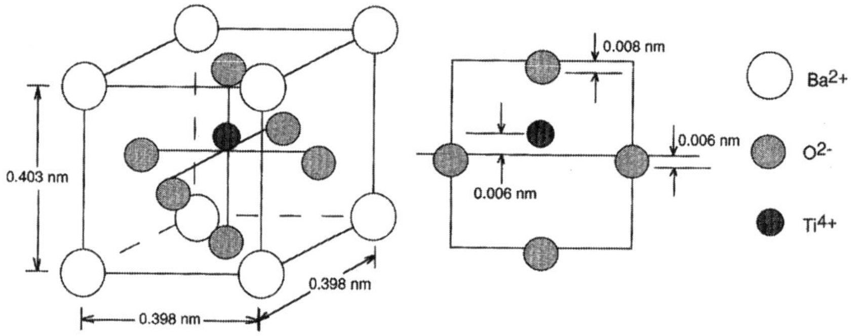

Fig. 4.16 The unit cell of barium titanate. Note the centers of negative and positive charges are separated, creating a permanent dipole. Source: Ref 4.1.

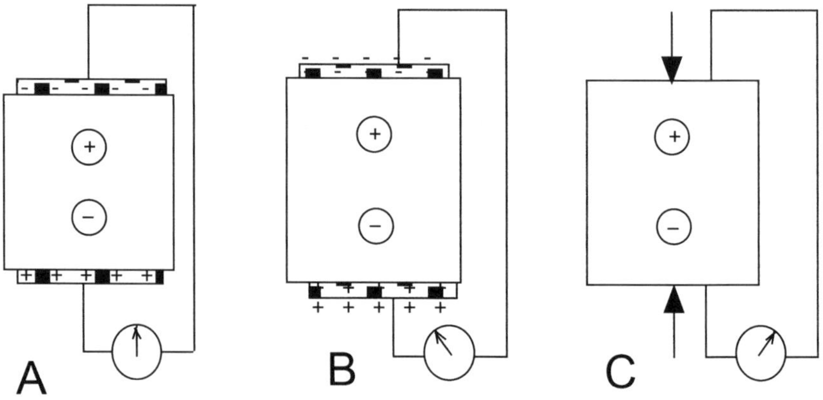

Fig. 4.17 A piezoelectric crystal (A) is distorted by an electric field (B) and an external force produces a voltage (C). Source: Ref 4.1.

NOTE OF INTEREST

Light-emitting diode (LED) traffic lights are brighter, use less energy, and have much longer lives than the old incandescent lights. However, they do have one disadvantage for use in northern states. They are so efficient that they do not emit enough heat to melt snow.

Vibrating quartz crystals often are used for frequency standards. Their natural vibrational frequency depends on the crystal dimensions. Once a crystal is cut to its final dimensions, its frequency is constant to one part in 10^8. Piezoelectric crystals are used as pressure sensors.

Thermal Conductivity

Metals have very high thermal conductivities. Most of the heat is transported by electrons, thus the thermal conductivity is closely related to the electrical conductivity. The ratio of thermal conductivity to electrical conductivity is about 7×10^{-6} w-Ohm/(K/m). This is called the Wiedeman-Franz ratio. In nonmetals, thermal conduction is primarily by lattice vibrations and correlates with the elastic modulus. Diamond, which has the highest elastic modulus, also has the highest thermal conductivity of all materials. Polymers have very low elastic moduli and as a result very poor thermal conductivity. This makes them useful for pot handles.

Exercise

1. **Conductors and Insulators**

   Using a multimeter to measure the resistance of several materials, determine which materials are conductors and which are insulators.

   You should find that all metals are conductors and almost everything else is an insulator.

REFERENCES

4.1 W.F. Hosford, *Materials for Engineers,* Cambridge U. Press, 2008.
4.2 L.H. Van Vlack, *Elements of Materials Science, 6th ed.,* Addison-Wesley, 1989.
4.3 W.F. Hosford, *Physical Metallurgy,* CRC, 2005.
4.4 A. Wert and R.M. Thompson, *Physics of Solids,* McGraw-Hill, 1970

CHAPTER 5

Magnetic Behavior

Ferromagnetism

Magnetism seems to be a mysterious phenomenon. The discovery by ancient people of lodestone (Fe_3O_4), a naturally magnetized piece of the mineral magnetite, led to many myths (Fig. 5.1). When "magnetic behavior" is spoken of, it usually means ferromagnetic behavior. Actually there are two *other* types of magnetic behavior: diamagnetic behavior, which is a weak *repulsion* of a magnetic field, and diamagnetism, which is a weak *attraction* of a magnetic field.

Ferromagnetism, in contrast, is a very strong attraction of a magnetic field. There are only a few ferromagnetic elements, the important ones being iron, nickel, and cobalt. Several rare-earth elements are ferromagnetic at low temperatures. Table 5.1 lists all of the ferromagnetic elements and the temperature above which they cease to be ferromagnetic (*Curie temperature*).

Fig. 5.1 Typical legend about lodestone, the first permanent magnet known to man. Source: Ref 5.1

Table 5.1 Ferromagnetic elements and Curie temperatures

Metal	Curie temperature, °C (°F)
Iron	1044 (1910)
Cobalt	1121 (2050)
Nickel	358 (675)
Gadolinium	16 (60)
Terbium	–40 (–40)
Dysprosium	–181 (–295)

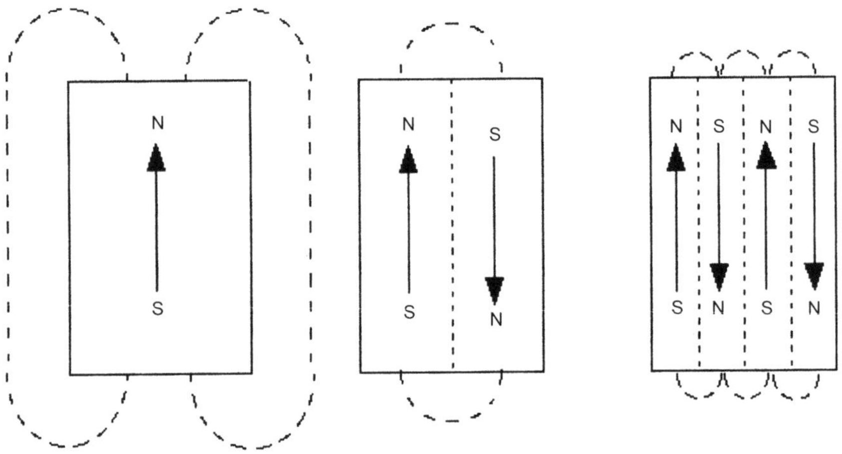

Fig. 5.2 Incomplete magnetostatic circuits raise the energy

Iron, nickel, and cobalt at room temperature always are magnetized, although they may not behave as magnets. The energy is lowered if the magnetic spin of each atom is closely aligned with that of its neighbor. This causes small regions, called domains, in which all of the spins are aligned. This lowering of energy by the alignment of spins is called the exchange energy.

Magnetostatic Circuits

There is a tendency of a material to form complete magnetic circuits, lowering the magnetostatic energy. The magnetostatic energy is less with smaller domains (Fig. 5.2). This is the reason why a horseshoe magnet attracts iron (Fig. 5.3). The net result is that domain structures are formed like those illustrated in Fig. 5.4.

If a piece of iron is placed in an external magnetic field, those domains most closely aligned with the field will tend to grow at the expense of those that are anti-aligned (Fig. 5.4).

Figure 5.5 is a graph showing how the net magnetization of a material, B, increases with the applied external magnetic field, H, over steps a., b., c., and d.

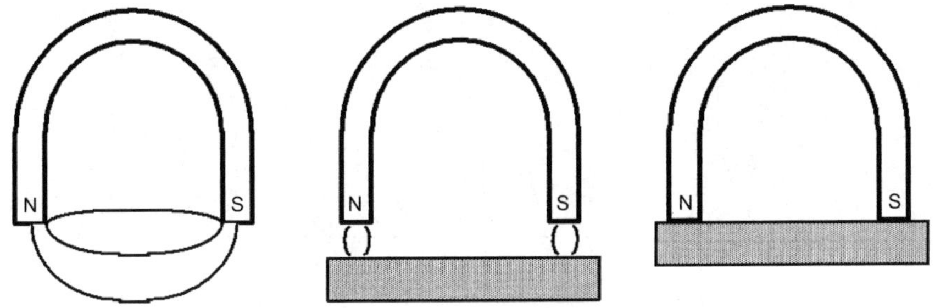

Fig. 5.3 Horseshoe magnets attract iron to complete magnetostatic circuits

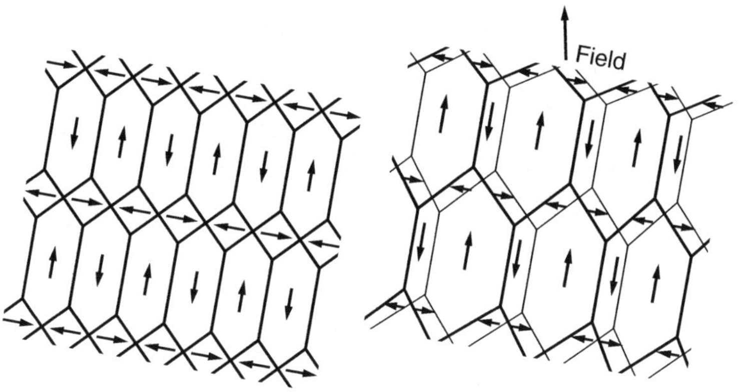

Fig. 5.4 Typical domain structures composed of complete magnetostatic circuits. Source: Ref 5.2

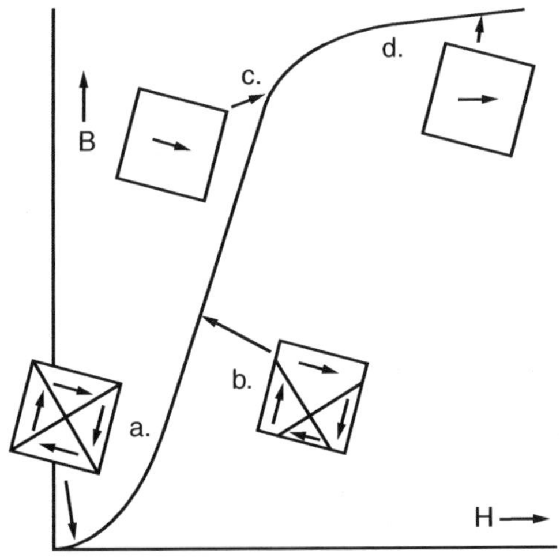

Fig. 5.5 Magnetization of a material. Source: Ref 5.1

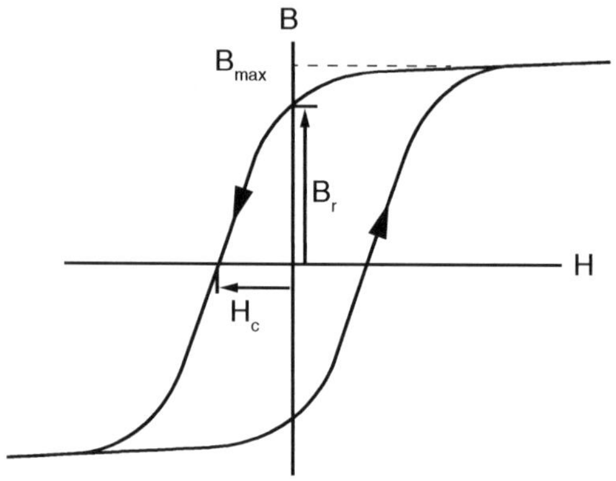

Fig. 5.6 A typical *B-H* curve. The magnetization initially increases by growth of favorably oriented domains. At high magnetic fields, the magnetization rotates out of the easy direction of magnetization toward the field. Source: Ref 5.1

A complete *B-H* curve is illustrated in Fig. 5.6. Initially, the magnetization increases with the applied magnetic field. A maximum magnetization, B_{max}, is reached when all of the spins are aligned with the external magnetic field. If the external field is removed, there will be a residual magnetization, B_r. A reverse field, H_c, called a *coercive force,* is necessary to demagnetize the material.

Domain Boundaries

The boundaries between domains are regions where there is a gradual change in the direction of magnetization. In a ferromagnetic material the direction of magnetization tends to be parallel to a specific crystallographic direction. Magnetization in any other direction raises the magnetocrystalline energy. The width of these (perhaps 20 atoms) is a compromise between the magnetocrystalline and exchange energy terms. A wider boundary would require more atoms to be magnetized in a direction other than the direction of easy magnetization. The exchange energy is minimized if the boundary is very wide, thus the direction of magnetization changes very little between neighboring atoms. There are two domain boundary possibilities in iron, 180° and 90° domain walls (Fig. 5.7).

Soft versus Hard Magnetic Materials

There are two main types of magnets: hard and soft. Hard magnetic materials are permanent magnets. They are difficult to demagnetize. The hysteresis is very large. The remanence, B_r, and coercive force, H_c, are high. Soft magnetic materials are easily demagnetized. The terms *soft* and *hard* are historic. The best permanent magnets in the 1910s were made of

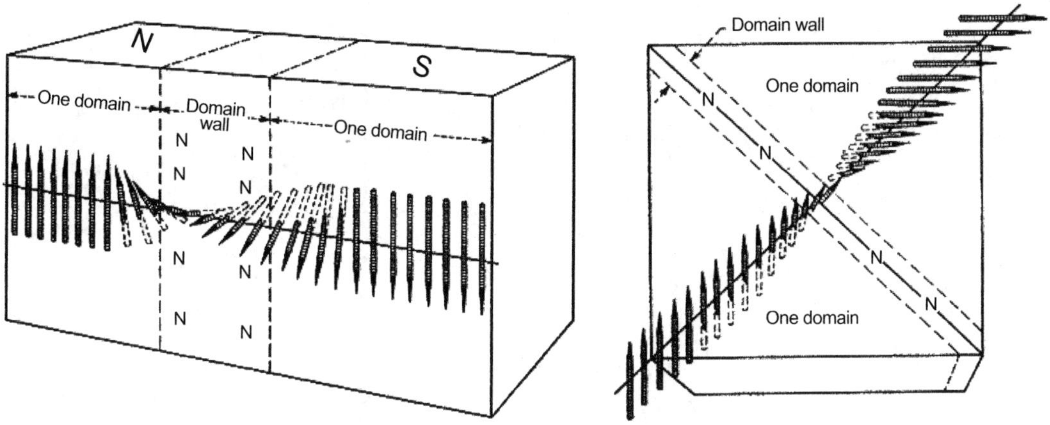

Fig. 5.7 Two domain boundary possibilities in iron, 180° and 90° domain walls. Source: Ref 5.1

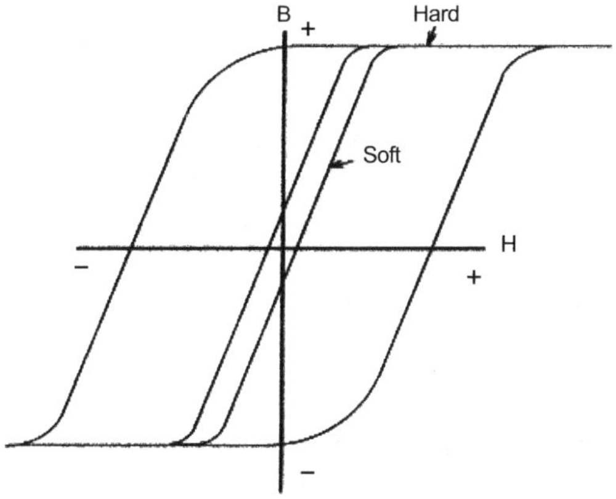

Fig. 5.8 A hard magnetic material has a much greater hysteresis than a soft magnetic material. The differences are much greater than shown in this figure. Source: Ref 5.2

martensitic steel, which is very hard, and the best soft magnets were made from pure annealed iron. The differences of the *B*-*H* curves are shown in Fig. 5.8. The values of H_c range from 0.002 oersteds for a superalloy to 4.300 for a platinum-cobalt alloy.

Soft Magnetic Materials

For a material to be soft magnetically, its domain walls must move easily. The principal obstacles to domain wall movement are inclusions and grain boundaries. Low dislocation contents and residual stresses, as well as a low interstitial content, also are important.

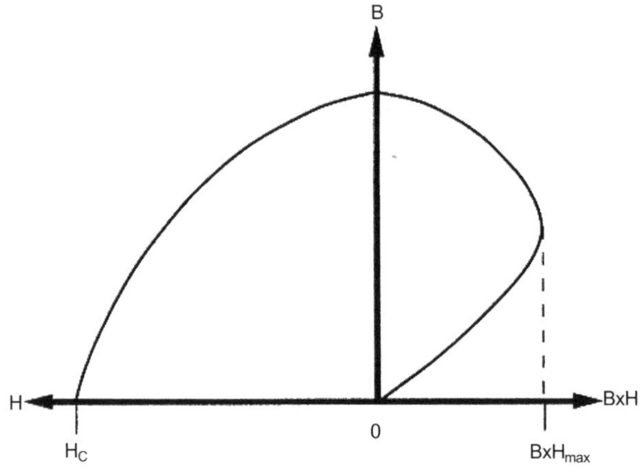

Fig. 5.9 Second quadrant of a *B-H* curve (left) and the corresponding $B \times H$ product (right). The maximum $B \times H$ product is a figure of merit. Source: Ref 5.2

Inclusions are important obstacles to domain wall movement because the energy of the system is lower when a domain wall passes through an inclusion than when the boundary has separated from the inclusion.

Uses of soft magnetic materials include transformers, motor and generator cores, solenoids, relays, magnetic shielding, and electromagnets for handling scrap. Many of these applications employ silicon-iron (usually 3 to 3.5% Si). Alloys containing ≥3% Si are body-centered cubic (bcc) ferrite at all temperatures up to the melting point (Fig. 5.9). Silicon increases the electrical resistance of iron. A high electrical resistance is desirable for transformers because eddy currents are one of the principal power losses in transformers. The power loss is inversely proportional to resistance, therefore thin sheets also minimize eddy currents.

Hard Magnetic Materials

A high H_c coercive force is desirable for hard magnets, but most important is a high $H \times B$ product in the second quadrant of the *B-H* curve (Fig. 5.9). Often the maximum $B \times H$ product is taken as a figure of merit.

Materials made up of small, isolated particles that are single domains make good hard magnets, particularly if the particles are elongated and if there is high magnetocrystalline energy. In a microstructure consisting of small isolated particles surrounded by a nonferromagnetic phase, there are no domain walls that can move. The direction of magnetization can be changed only by rotating the magnetization out of the easy direction into another equivalently easy direction. If there is a great dependence of the energy on the direction of magnetization, this will require a high magnetic

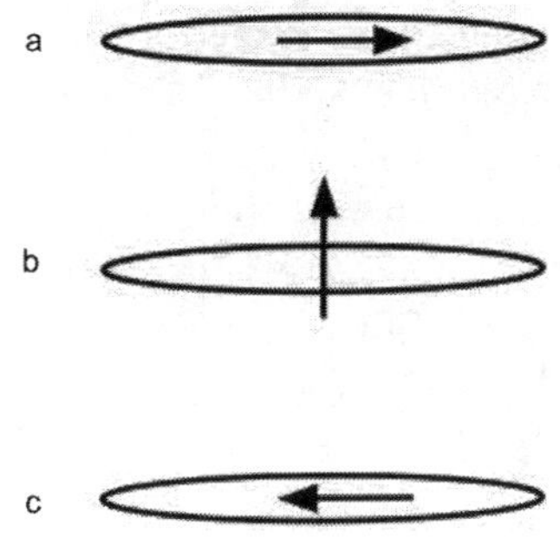

Fig. 5.10 As the direction of magnetization of an elongated particle is reversed (*a* vs. *c*), the rotation of the direction of magnetization increases the magnetostatic energy (*b*). Source: Ref 5.1

> **NOTE OF INTEREST**
>
> Lodestone is magnetite, Fe_3O_4, which is ferromagnetic. It formed the basis for the first compasses. In the 4th century B.C. the Chinese recorded that iron was attracted to lodestone. The Chinese used compasses with lodestone as early as 1100 A.D. The compass was used by western Europeans in the late 1100s and by Arabs in the early 1200s. The first real use of the magnetic phenomenon likely should be attributed to the Vikings. Their development of magnetic compasses enabled them to travel far at sea, even in foggy conditions.

field. Hexagonal structures are useful here because there are only two easy directions, both parallel to the hexagonal axis and differing by 180°. When ferromagnetic particles are elongated, the intermediate stage will have a high energy (Fig. 5.10).

Inexpensive permanent magnets can be made by aligning fine iron powder in a magnetic field while it is being bonded by rubber or a polymer.

Exercise

1. **Magnetic Domains**

   The following will allow observation of magnetic domains. Make a very fine slurry of magnetite using the following recipe:

   i. Dissolve 2 g of ferrous chloride ($Fe_2 \cdot 4H_2O$) and 5.4 g of ferric chloride ($FeCl_3 \cdot 6H_2O$) in 300 cc of hot water in a 500 cc beaker.

   ii. While stirring, add a solution of 5 g of sodium hydroxide dissolved in 50 cc of water. This will form a black precipitate.

   iii. After the precipitate has settled to the bottom, pour off the clear liquid.

iv. Transfer the remainder to a 2-liter beaker and fill with clean water. After the precipitate has settled to the bottom, pour off the clear liquid again.
v. Prepare a stabilizing solution by placing 5 g of coconut oil amine (mostly dodecylamine) into a 1-liter beaker, adding 50 cc of water and gently warming until the amine has melted. This must be neutralized by adding HCl drop by drop until litmus paper indicates it is neutral. Next add 300 cc of warm water and combine with the precipitate, then thoroughly mix in a blender.
vi. Finally, obtain a small piece of silicon-iron sheet (approximately 9.5 mm, or 3/8 in., square). Polish the sheet to approximately 0.05 mm (0.002 in.) thickness, put in a microscope, and place a small amount of the solution on it. Observe motion of the domains as a magnet is placed in the vicinity and moved.

Movement of domain boundaries should be visible.

REFERENCES

5.1 E.A. Nesbitt, *Ferromagnetic Domains,* Bell Telephone Labs, 1962
5.2 W.F. Hosford, *Physical Metallurgy,* 2nd ed., CRC Press, 2010

CHAPTER 6

Nonferrous Metals

METALS ARE EASILY recognized by their luster. They are characterized by their high electrical and thermal conductivities. These properties can be understood in terms of the nature of metallic bonding. In metals, positively charged atoms are surrounded by a cloud of valence electrons that are free to move about the lattice, accounting for high electrical conductivity. Metals have high thermal conductivities because the free movement of electrons allows them to transfer heat. Electrical conductivity decreases with anything that disturbs the periodicity of the lattice, including high temperatures. Thermal vibrations cause atoms to be displaced from their normal positions. Atoms in solid solution have the same effect. Because the valence electrons are mobile, they can absorb energy from photons, increasing their energy but immediately dropping back to a lower energy state and emitting photons. This explains the metallic luster.

Most metals are strong and ductile. They deform plastically because their atomic planes slip over one another, as illustrated in Fig. 6.1. Metals

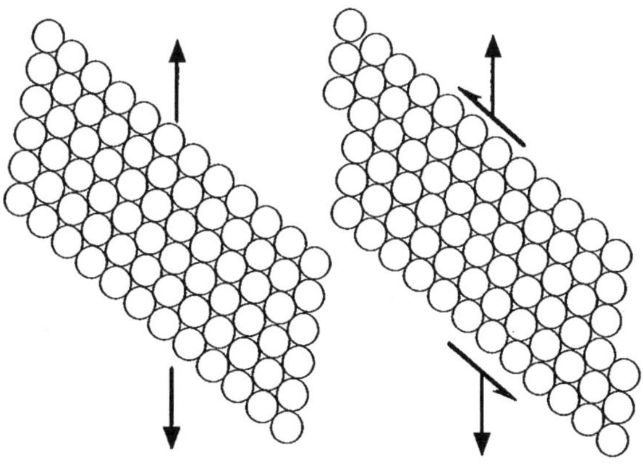

Fig. 6.1 Deformation of crystals by slip. One plane of atoms slides over another.

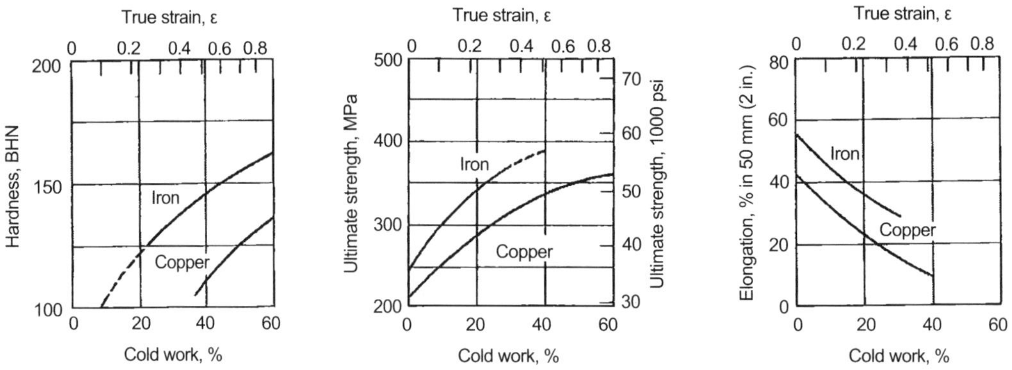

Fig. 6.2 Effects of cold work on the hardness, tensile strength, and ductility of copper and iron. Here cold work means the percent reduction of thickness by rolling. Source: Ref 6.1

such as copper, aluminum, iron, titanium, magnesium, and nickel are very ductile. Other metals, including bismuth, manganese, and bronzes, are brittle and break before they deform very much.

Cold Work

Cold working, which alters the shape of a metal, also strengthens it by plastic deformation. Plastic deformation leaves defects in crystals, thus making continued slip more difficult. This is the reason the stress-strain curves in tension show stress rising with strain. Straining a metal, whether in a tension test or by rolling or drawing it into thinner sections (in cold work), causes its hardness, yield strength, and tensile strength to increase, and its remaining ductility to decrease (Fig. 6.2).

Annealing

When a cold worked material is annealed, or put through a process in which it is heated and then cooled, changes in properties and structure occur that can be divided into three stages: recovery, recrystallization, and grain growth.

During the *recovery* stage there is a rearrangement and decrease in the number of defects in the metal. Lattice defects generated during cold working lower the electrical conductivity, but recovery largely restores the electrical properties. During recovery there also is a significant decrease in the residual stresses that were caused by cold working. The hardness does not change very much during recovery. The effects of recovery in tungsten are apparent in Fig. 6.3.

At higher annealing temperatures new grains start to form that contain fewer defects and are much softer. This process, called *recrystallization,* is accompanied by a marked decrease in hardness and an increase in ductil-

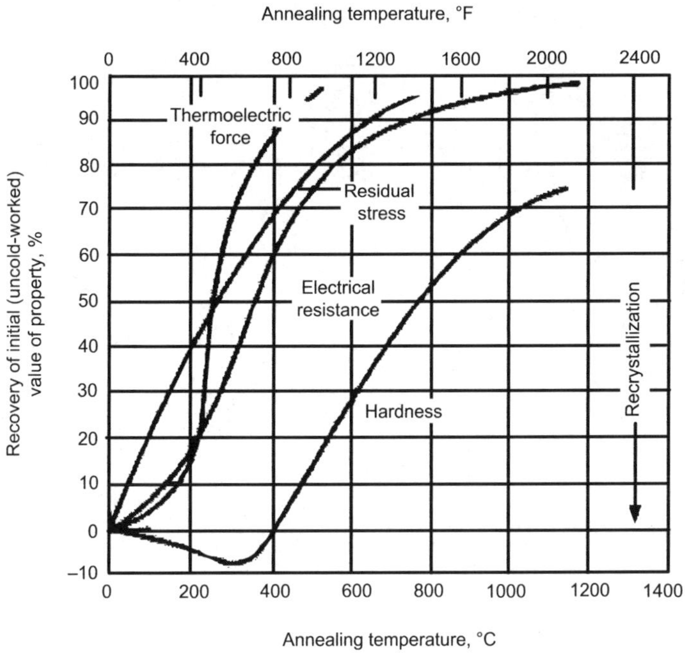

Fig. 6.3 Property changes during annealing of tungsten. Only recovery occurs at temperatures below approximately 400 °C (700 °F). Source: Ref 6.2

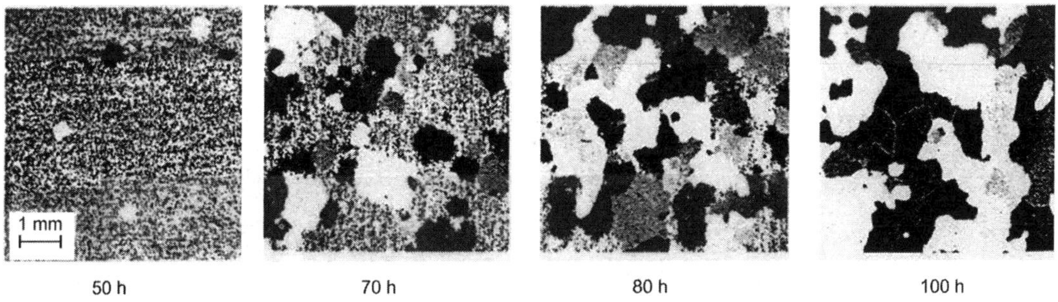

Fig. 6.4 Progress of recrystallization at 310 °C (590 °F) of aluminum that had been cold worked 5%. Note that a few new grains are present after 50 h, and these grow bigger as new ones are nucleated, after 70–80 h. After 100 h, recrystallization is complete. Source: Ref 6.3

ity. Figure 6.4 shows new grains forming and growing into old deformed grains over time.

Figure 6.5 shows there is a large decrease in hardness during recrystallization. It also shows that increasing the amount of cold work tends to lower the temperature at which recrystallization occurs. Impurities present as second phases also retard the movement of grain boundaries, thereby raising the recrystallization temperature.

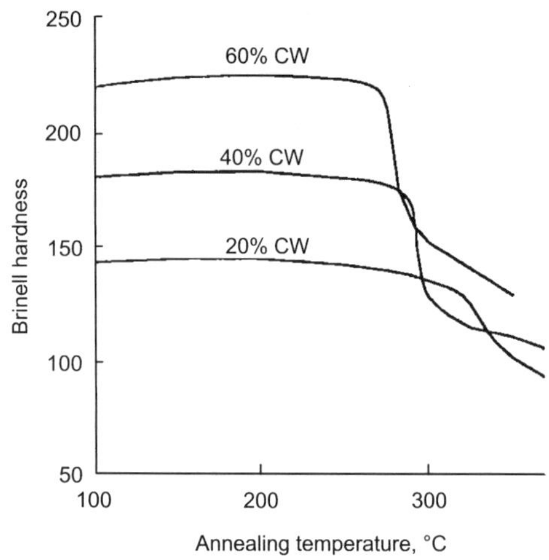

Fig. 6.5 Effect of annealing temperature on cold worked (CW) brass. Note that the sharp drop in hardness associated with recrystallization occurs at lower temperatures for material that is more heavily cold worked. Source: Ref 6.1

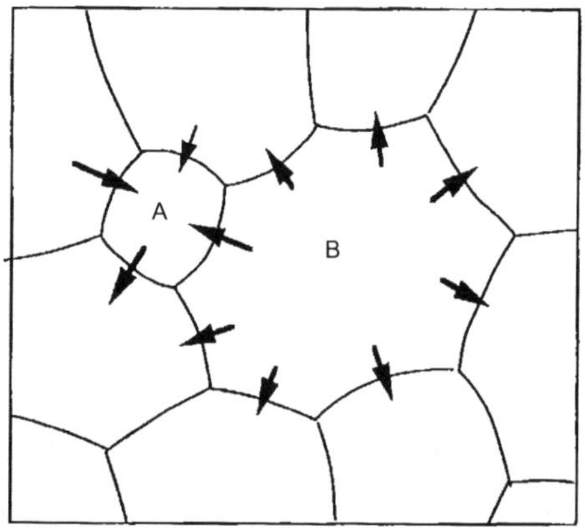

Fig. 6.6 Large grains tend to be outwardly concave, and, therefore, grow at the expense of their neighbors. Small grains tend to be inwardly concave and therefore shrink. Source: Ref 6.4

Grain growth, the third stage of annealing, involves the growth of some recrystallized grains. The equilibrium angle between grain boundaries is 60°; thus, on average, each grain has six near neighbors. Larger grains statistically have more neighbors than small grains, thus the grain boundaries on large grains are outwardly concave (Fig. 6.6). Grain boundaries tend to shorten themselves by growing toward their center of curvature. Thus large grains grow at the expense of smaller grains. A grain that is smaller than its neighbors has less than six neighbors, so its boundaries are inwardly concave, and it will shrink until it disappears. The result will be fewer grains, thus the average grain size will increase.

There is some continued decrease in hardness during grain growth. This is apparent in Fig. 6.5 for the 60% cold worked material above 280 °C (540 °F), the 40% cold worked material above 300 °C (600 °F), and the 20% cold worked material above approximately 340 °C (640 °F).

While all metals share important characteristics, metals can be divided into two main categories: ferrous and nonferrous, the two groups categorized by the presence and absence of iron, respectively. The remainder of this chapter discusses properties, behavior, and uses of several nonferrous metals.

Copper

Copper has very high electrical conductivity and therefore it is widely used for wiring. All alloying additions lower the conductivity of copper. Copper alloys such as brass, which is an alloy with zinc, and bronze, which is an alloy with tin, are much stronger than pure copper. This is because the different size atoms interfere with slip. Brass contains up to 35% Zn and bronze approximately 5% Sn. Both alloys are often used because of their color and because of their corrosion resistance in many environments.

Aluminum

The modern process for producing aluminum metal from aluminum oxide was discovered independently and almost simultaneously in 1886—in Ohio by Charles Martin Hall and in France by Paul Héroult. Hall opened the first large-scale aluminum plant in 1888. Until that time, aluminum was very expensive. In fact, Napoleon had dinnerware made of aluminum because its expense made it very stylish.

Aluminum also has very high electrical conductivity. It is used for power transmission because it is less expensive than copper. Many of the uses of aluminum are due to its low density. These include transportation (bodies, trim, and other parts of autos, trucks, and aircraft). Other uses include packaging (beverage cans and foil) and building and construction (siding, roofing, doors, and windows). Its low electrical resistance makes it useful in electrical wiring and overhead transmission lines, while corrosion resistance makes it useful for cooking utensils. High reflectivity is responsible for its use in insulation and mirrors. Figure 6.7 indicates relative uses of aluminum.

An aluminum alloy containing approximately 4% Cu (Fig. 6.8) can be hardened by a special heat treatment called *precipitation* or *age hardening*. This involves heating the alloy to just under its eutectic temperature to dissolve the copper into the aluminum lattice, followed by quenching

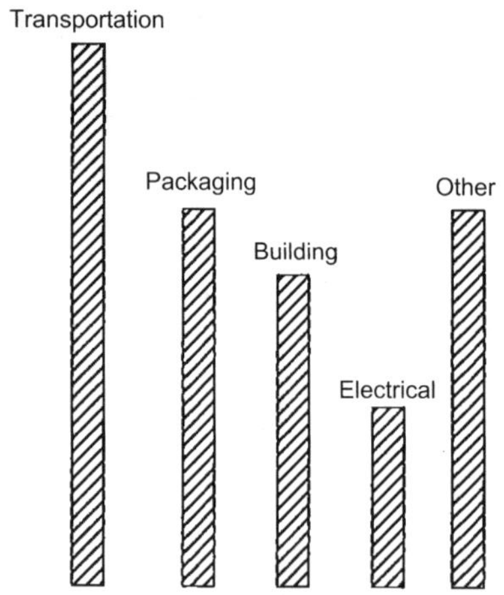

Fig. 6.7 Relative uses of aluminum. Packaging includes beverage cans. Transportation includes cars, trucks, and aircraft.

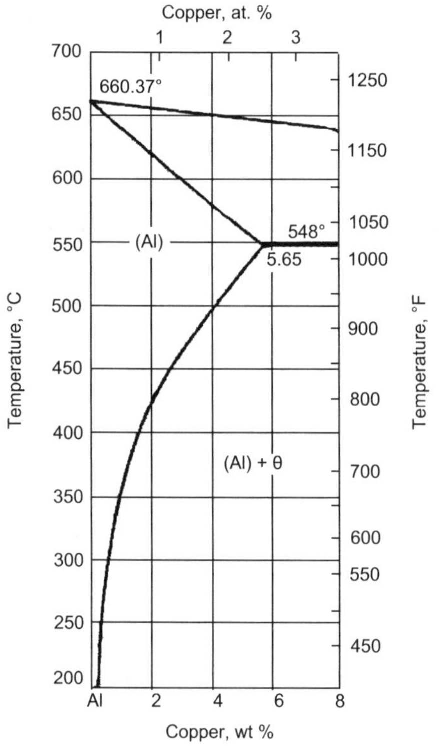

Fig. 6.8 The aluminum-rich end of the aluminum-copper phase diagram. An alloy containing 4% Cu can be hardened by heating to 530 °C (990 °F) to dissolve the copper, quenching to room temperature to form a supersaturated solid solution, and finally heating to approximately 200 °C (400 °F) to allow fine precipitates to form. Source: Ref 6.5

to room temperature. This supersaturated solid solution then is reheated to a lower temperature, at which a copper-rich phase can precipitate. If the precipitation temperature is properly chosen, the precipitate particles will be very small, hardening the material by blocking slip. Other precipitation-hardenable aluminum alloys contain either zinc or Mg_2Si in solution. Other alloys can be strengthened by precipitation hardening. In fact, even nonmetallic systems are hardened by fine dispersions of hard particles. Examples include hardening of asphalt by sand, and Bakelite by fine silica particles.

Most aluminum castings contain silicon because there is little volume change on freezing due to the expansion of silicon virtually cancelling out the shrinkage of aluminum. Making sound castings is simplified if there is little or no shrinkage when liquid metal freezes.

Magnesium

Magnesium is even lighter than aluminum. Because of its low density, magnesium is finding increasing use in aircraft and automobiles. Because finely divided magnesium was used in incendiary bombs, magnesium suffered from an unjustifiable fear of fire hazard. It was used for many years as engine blocks and transmission housings in early Volkswagen automobiles without any resulting fires. Some magnesium alloys are precipitation hardenable. Its room-temperature ductility is limited; thus, most of its applications are as castings. Other uses include laptop computer cases and lightweight tool casings.

Titanium

Titanium has a hexagonal close packed (hcp) crystal structure at room temperature, but titanium transforms to a body-centered cubic (bcc) structure above 883 °C (1620 °F). Alloying elements either lower or raise this temperature. This transformation allows heat treatments to be used to harden some titanium alloys.

The density of titanium is between those of aluminum and steel. It is stronger than aluminum alloys, thus it offers weight savings over steel where high strength is required.

The low density of titanium accounts for the majority of its usage. The aircraft industry uses titanium for airframes and engines. Its low density also is useful in helmets and armored vests.

A very tenacious surface oxide forms on titanium, giving it excellent corrosion resistance. Its resistance to corrosion in saltwater is useful in power plants using salt water, in desalinization plants, and wherever seawater corrosion is a problem. Because titanium is inert to human body fluids, it is an ideal metal for medical replacement structures such as hip and knee implants. Titanium actually allows bone growth to adhere to the implants, thus titanium implants are more durable than those made of other materials.

Zinc

The only major use of zinc-base alloys is for small parts such as locks, and small automotive parts made by die casting. Die casting is a process in which liquid metal is forced into a water-cooled metal mold and automatically ejected when it is adequately cool. It is used when a very large number of parts are to be made. The largest use of zinc is in galvanizing, which is plating with zinc. Steel sheets often are galvanized with zinc for corrosion protection. Another major use is for alloying with copper to form brass.

Lead and Tin

These low-melting metals were used in the past for solder, but because of the health hazard of lead, modern solders contain no lead. The major use of lead is in automobile batteries. In the past tin was used to plate steel cans. However, the high price of tin has almost eliminated this use. It is used as an alloying element in copper and titanium alloys.

NOTE OF INTEREST

Aluminum was chosen for the pyramidal cap of the Washington Monument. This was many years before the Hall-Heroult process that made aluminum relatively inexpensive, so aluminum was very expensive at the time. Aluminum was not the first choice for the cap. It was selected as an alternative material to copper-base alloys during discussions between Colonel Casey, the engineer in charge of the project, and William Frishmuth, who was the only U.S. aluminum producer at that time. Casey sent a request to Frishmuth asking if he could make a metal pyramidal cap that was to serve as the lightning rod of the monument. Copper, bronze, and platinum-plated brass were the preferred materials. Frishmuth proposed that the cap be made of aluminum at a quoted price of $75, and if he was unsuccessful in casting it in aluminum, he would cast it in aluminum bronze and plate it with gold for a total cost of $50, or, if platinum-plated, for $75. Casey, in accepting Frishmuth's proposal, admonished him to keep the costs within the estimate if possible. Frishmuth's bill for the aluminum pyramid was for $256.10. When the bill arrived, Casey sent his assistant, Captain Davis, to go to Philadelphia to investigate the matter. After Davis concluded his audit, a final price of $225 was agreed upon between Casey and Frishmuth.

Exercises

1. **Annealing Aluminum**

   Take two pieces of aluminum and bend them back and forth, and note how they become harder to bend. Then heat one in a flame. After it is cool, bend it and note that heating softened it.

   This is a result of annealing.

2. **Dispersion Hardening**

   Put a handful of peanuts into a jar of peanut butter and note how much stiffer the peanut butter is than before the peanuts were added.

   This is an example of dispersion hardening.

3. **Heat Reflection**

   Place the bulbs of thermometers into two equal weights of food and wrap each tightly with aluminum foil. Place the shiny side of the foil out on one, and the dull side out on the other. Heat them together in an oven. Which heats faster? Why?

   It should be found that the one with the dull side out heats faster. This illustrates that shiny aluminum reflects heat.

REFERENCES

6.1 L.H. Van Vlack, *Elements of Materials Science,* 6th ed., Addison-Wesley, 1989
6.2 A. Guy and J. Hren, *Elements of Physical Metallurgy,* 3rd ed., Addison-Wesley, 1974
6.3 W.C. Leslie, J.T. Michalak, and F.W. Auk, *Iron and its Dilute Solutions,* AIME Conference Series, Interscience, 1963
6.4 W.F. Hosford, *Physical Metallurgy,* 2nd ed., CRC Press, 2010
6.5 *Metallography, Structures and Phase Diagrams,* Vol 8, *Metals Handbook,* 8th ed., American Society for Metals, 1973

CHAPTER 7

Iron and Steel

WHILE THE LAST CHAPTER dealt with nonferrous metals, this chapter focuses on the other main category of metals: ferrous metals, or those containing iron.

Iron is the most versatile of all metals. It is inexpensive and can be made very hard or relatively soft and ductile depending on carbon content, alloying elements, and heat treatment. Its crystal structure at room temperature is body-centered cubic (bcc). Pure iron transforms to a face-centered cubic (fcc) structure above 910 °C (1670 °F). Most alloying elements lower this transformation temperature.

Steels

Most alloys of iron are called steels. Carbon is the most important alloying element in steel, although not all steels contain carbon. The solubility of carbon in the fcc form of iron is very much greater than its solubility in the bcc form. Figure 7.1 shows the iron-rich end of the iron-carbon phase diagram. The α and δ phases are bcc and called *ferrite*. The γ phase is fcc and called *austenite*. Iron carbide (Fe_3C), also called *cementite,* forms from austenite during cooling.

If iron containing 0.77% C is cooled slowly through the eutectoid temperature of 730 °C (1340 °F), the eutectoid reaction, $\gamma \rightarrow \alpha + Fe_3C$, occurs with both the ferrite (α) and the cementite (Fe_3C) precipitating simultaneously in the form of alternating platelets (Fig. 7.2). This microstructure is called *pearlite*.

When a steel that contains less than the eutectoid composition of 0.77% C is slowly cooled, ferrite will form before the eutectoid temperature is reached. As the ferrite forms, the austenite becomes enriched in carbon, finally reaching 0.77% C at the eutectoid temperature. With further cooling, the austenite transforms to pearlite. Figure 7.3 shows the resulting microstructure with ferrite in the former austenite grain boundaries. In steels with more than 0.77% C, cementite will form in the austenite grain boundaries before pearlite forms.

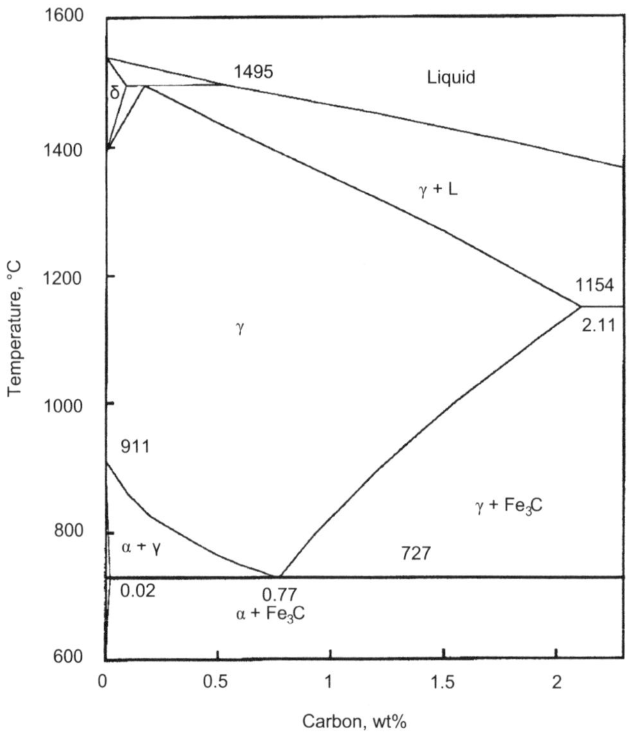

Fig. 7.1 The iron-rich end of the iron-carbon phase diagram. The phase region labeled γ is face-centered cubic and the phase regions labeled α and δ are body-centered cubic. Iron carbide (Fe_3C) contains 6.67% C. Source: Adapted from Ref 7.1

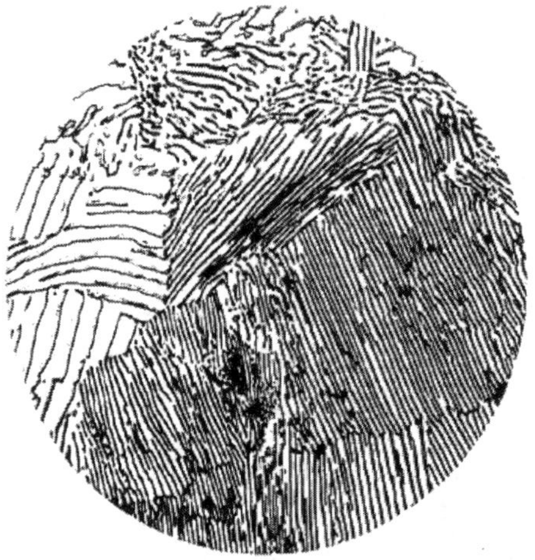

Fig. 7.2 Microstructure of pearlite. Note the alternating platelets of ferrite (white) and iron carbide (dark). Source: Ref 7.2

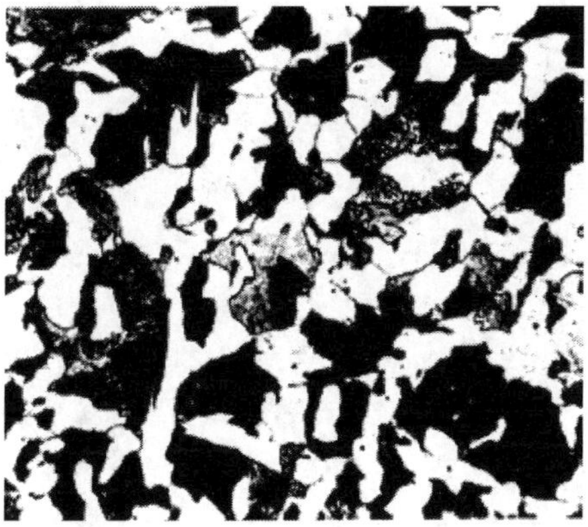

Fig. 7.3 Microstructure of a steel containing 0.45% C after air cooling. The light regions are ferrite, which formed in the prior austenite grain boundaries. The black areas are pearlite, in which the spacing between lamellae is too small to be resolved. Source: Ref 7.3

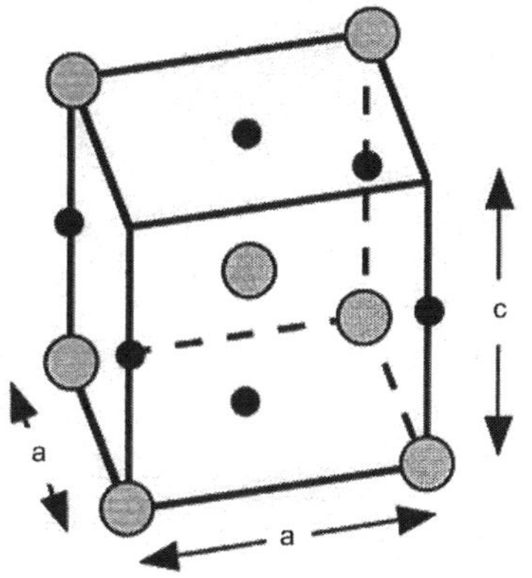

Fig. 7.4 The martensite unit cell. Ref 7.4

Hardening

If austenite is rapidly cooled, the carbon atoms do not have enough time to diffuse away from the growing ferrite to the growing cementite. Instead, a new phase, *martensite,* will form. Martensite essentially is a supersaturated ferrite. Its crystal structure is body-centered tetragonal, or bct (Fig. 7.4), which is like a distorted bcc structure. The amount of distortion depends on the amount of carbon in the steel. Figure 7.5 shows how the ratio of the edges of the unit cell depends on the carbon content. With no carbon the unit cell is bcc.

Martensite is harder than pearlite. Its hardness depends only on the carbon content (Fig. 7.6).

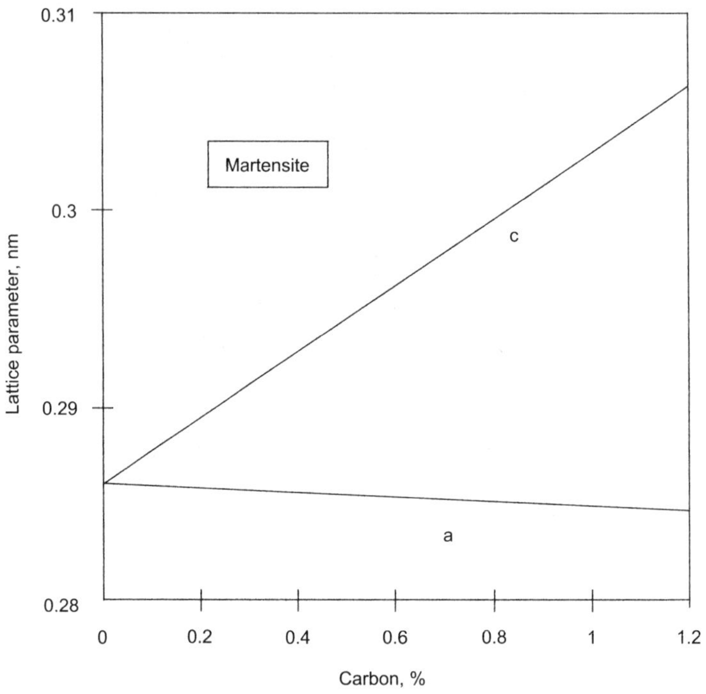

Fig. 7.5 The ratio of the c lattice parameter to the *a* lattice parameter increases with carbon content. With no carbon, the lattice parameters are the same so the structure is cubic. Source: Ref 7.4

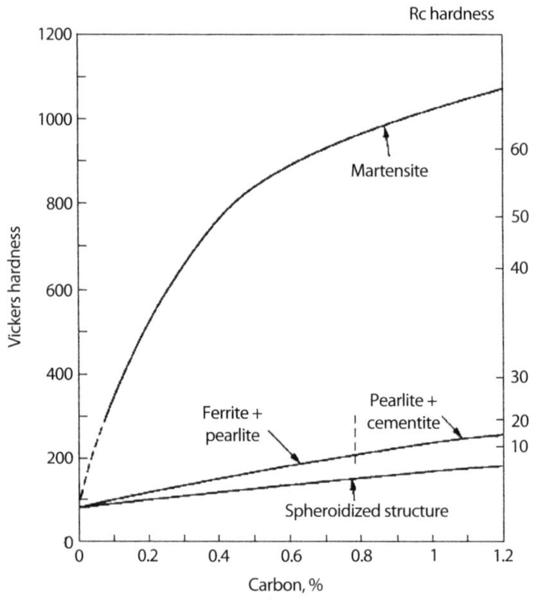

Fig. 7.6 The hardness of martensite increases with carbon content. Note that the hardness extrapolates to the hardness of ferrite at 0% C. The hardnesses of pearlitic and spheroidized structures are shown for comparison. Source: Ref 7.5

Alloying elements in austenite slow the transformation to pearlite, because they must diffuse to either the growing cementite or ferrite platelets. Neither nickel nor silicon form carbides in steel and, therefore, segregate to the ferrite. Manganese, chromium, molybdenum, and tungsten are carbide formers, so they segregate to the cementite. In both cases, the diffusion of these elements in substitutional solid solution is significantly slower than the diffusion of carbon, and, therefore, they slow the transformation of austenite to pearlite.

If parts made from plain carbon steels are too thick, their centers may not cool fast enough to avoid pearlite formation, and they have insufficient *hardenability*. Alloying elements increase the hardenability by allowing the steels to harden under somewhat slower cooling rates.

The Jominy end-quench test is a simple way to assess the hardenability of a steel. Water is sprayed on the end of a bar that has been heated to form austenite (Fig. 7.7). This creates a gradient of cooling rates along the bar, with the quenched end cooling very rapidly. Hardness measurements then are made along the length of the bar. Figure 7.8 shows typical results of the Jominy end-quench test for several steels. The compositions of these steels are:

	Composition, %				
Steel	**C**	**Mn**	**Cr**	**Mo**	**Ni**
4340	0.40	0.75	0.77	0.30	1.71
4140	0.38	0.79	1.01	0.22	...
1060	0.62	1.81	...	...	...
1040	0.39	0.89	...	...	...
1020	0.20	0.90	...	...	...

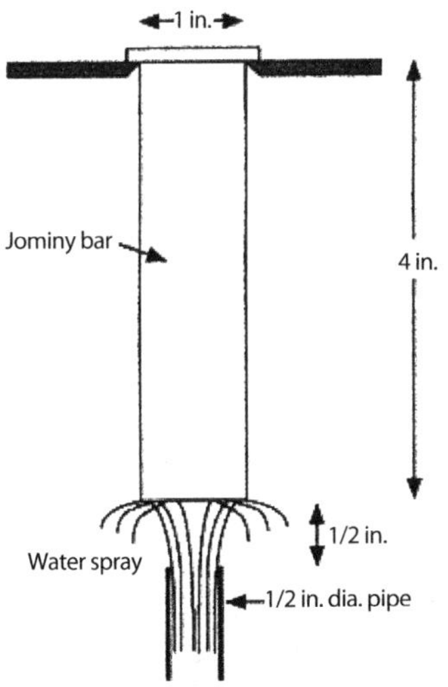

Fig. 7.7 Jominy end-quench test. Source: Ref 7.4

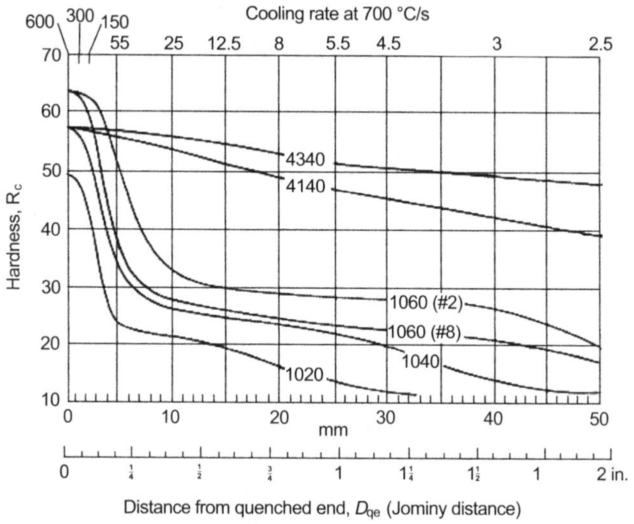

Fig. 7.8 Jominy end-quench test curves for several steels. Source: Ref 7.5

The 1060(#2) steel in Fig. 7.8 has a larger austenite grain size than the 1060(#8) steel. Several trends are apparent. The hardness at the quenched end where the structure is 100% martensite depends only on carbon content. Alloying elements have no effect on the hardness of martensite. The depth of hardening (hardenability) increases with alloy content, with carbon content, and with larger austenite grain size. The grain size effect is explained by preferential nucleation of pearlite in austenite grain boundaries. With a larger austenite grain size, nucleation of pearlite is slower, allowing more time for martensite to form. The principal use of alloying elements in steels is to increase their hardenability.

Tempering

Martensite is too brittle for most applications. A hardened steel usually is heated to some temperature considerably lower than the eutectoid temperature to *temper* it. Tempering is the conversion of hard brittle martensite into products that are more ductile but not as hard as as-quenched martensite. The amount of tempering increases with higher temperatures (Fig. 7.9). More tempering is required for hammers, which must be able to sustain impact loading, than for knives, which must be hard to remain sharp. Razor blades are tempered even less.

Special steels are used for applications in which a high hardness must be maintained at a high temperature, such as tools for high-speed machining. High contents of chromium, molybdenum, and tungsten in high-speed steels form very hard carbide precipitates when tempered at a high temperature. Figure 7.10 shows the effect of molybdenum on the

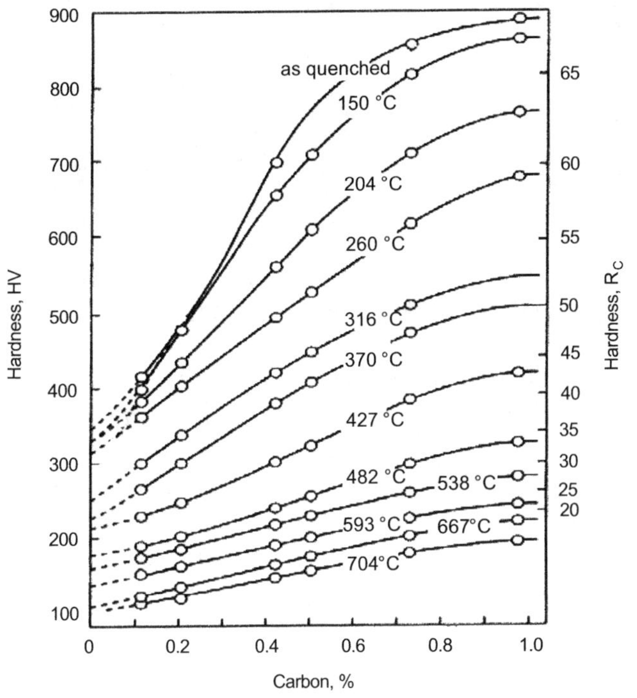

Fig. 7.9 Hardness of plain carbon steels after tempering at various temperatures for 30 min. Source: Ref 7.4

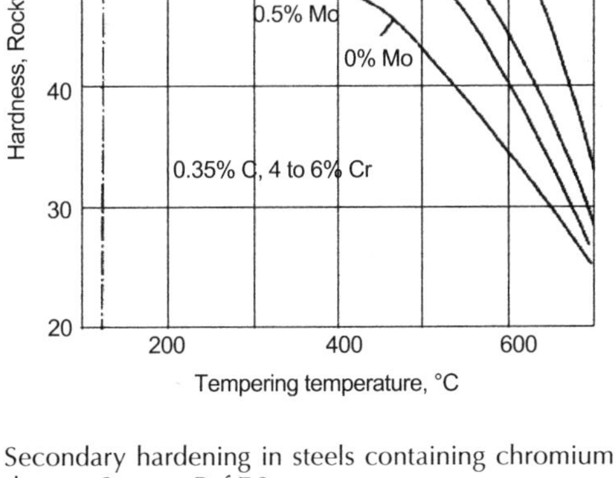

Fig. 7.10 Secondary hardening in steels containing chromium and molybdenum. Source: Ref 7.2

tempering response. With 5% Mo, high hardnesses are maintained at 550 °C (1020 °F).

Low-Carbon Steels

Most steels contain 0.10% C or less with no alloying elements except some manganese and silicon. These steels are classified as either *hot rolled* or *cold rolled* steels. Hot rolled grades are somewhat less expensive because they are finished hot but they have rough surfaces. They are used where surface appearance is not important, such as for I-beams and structural plate.

Cold rolled steels are first hot rolled, pickled in acid to remove scale, then cold rolled, and finally annealed in an inert atmosphere. Automobile bodies, household appliances, and other consumer items are made from cold rolled steel because of their very smooth surfaces. Formability may be of concern if the deformation during stamping is severe. For that reason, the carbon content of steels to be cold formed usually is less than 0.06% and in some cases is as low as 0.002%. In recent decades, concern over weight has led to the introduction of steels with somewhat higher yield strengths. *High-strength, low-alloy (HSLA) steels* are strengthened by a combination of precipitation of carbonitrides, finer grain sizes, and solid-solution effects. The formability of these steels is somewhat lower than plain carbon steels of the same carbon content, but adequate for most automobile parts. Dual-phase steels, which are quenched from the ferrite-austenite region, contain primarily ferrite with some martensite.

Stainless Steels

Alloying elements have very little effect on the corrosion behavior of steels. The exception to this is steels having a chromium content of at least 12%. Such *stainless steels* are very resistant to corrosion in the presence of oxygen. There are three main types of stainless steels:

- *Ferritic stainless steels* contain 12 to 18% Cr and very little carbon (generally less than 0.03%). Even when heated to temperatures just below their melting points, they maintain a bcc crystal structure.
- *Austenitic stainless steels* contain enough nickel that they have a fcc crystal structure at all temperatures below melting. A typical composition is 18% Cr, 8% Ni with very low carbon. They are more expensive than ferritic stainless steels and are nonmagnetic.
- *Martensitic stainless steels* have compositions similar to ferritic stainless steels but contain enough carbon so that when heated they transform to austenite. The hardenability is so high that they form martensite even with slow cooling. Applications include razor and knife blades.

Cast Irons

Cast irons contain far more carbon than steel. Typical carbon contents range from 2 to 4% C with 1 to 3% Si. With so much carbon and silicon, graphite, rather than cementite, tends to form on freezing. With the cooling rates typical of casting into sand molds, the graphite forms as flakes. Figure 7.11 shows a typical microstructure of a *gray cast iron*. The ductility is limited to a few percent elongation because fracture occurs easily along the graphite flakes. The name gray cast iron comes from the appearance of the fracture surfaces, which are almost entirely graphite.

If the sulfur content is kept very low, and magnesium or cerium is added to the melt just before casting, the graphite will form as spheroids. The resulting product is called *ductile cast iron*. Figure 7.12 shows a typical microstructure. The mechanical behavior of ductile cast iron is similar to that of steel.

Both gray and ductile cast irons may be reheated to form a structure of graphite and austenite. The rate of cooling, then, determines whether the matrix is pearlite, ferrite, or a mixture of pearlite and ferrite. Rapid cooling can be used to form martensite. Austempering of ductile cast iron produces many useful products. Gray and ductile cast irons are less expensive than cast steel because they melt at much lower temperatures than steels and are easier to cast because there is very little shrinkage on freezing.

White cast iron is a third grade of cast iron. The carbon and silicon contents of white cast iron are lower than in gray and ductile cast irons, and the cooling rates are greater. These factors tend to prevent graphite formation; thus, carbon occurs as iron carbide (cementite). White cast iron

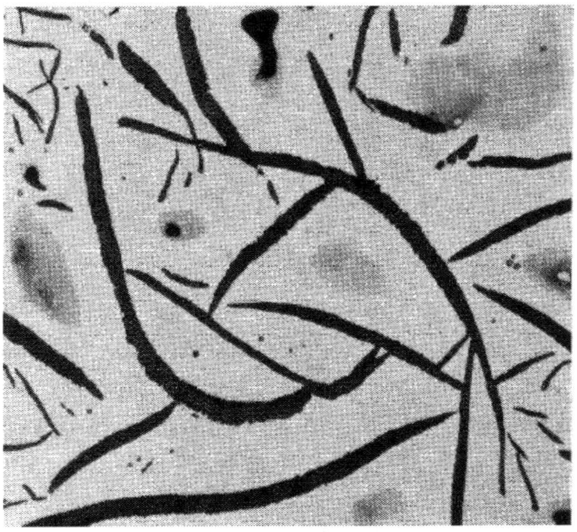

Fig. 7.11 Microstructure of gray cast iron. The black flakes are graphite, the white areas are ferrite, and the grey areas are pearlite. Source: Ref 7.6

NOTE OF INTEREST

When in about 2000 B.C. the Hittites in Asia Minor discovered how to reduce iron ore and make iron weapons, they were able to conquer the Babylonians, who had a civilization superior to their own, and to drive the Egyptians out of Syria. Their success was due to having harder and more weapons. Iron was far more plentiful than the copper and tin used for bronze weapons.

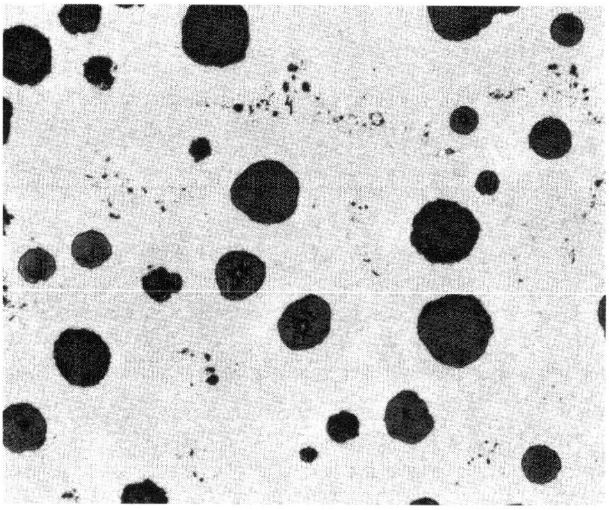

Fig. 7.12 Ductile cast iron. The spheroids are graphite and the white areas are ferrite. Source: Ref 7.6

is very hard and brittle and finds applications only where surfaces must be hard to resist wear.

Exercises

1. **Density of Austenite versus Ferrite**

   Stretch a piece of music wire (piano wire) between two fixed posts and heat it by passing a current through it. As it is heated it should sag because its length is changing. However, when it is heated through the ferrite-to-austenite transformation, it should rise suddenly.

   The rise occurs because austenite is denser than ferrite.

2. **Graphite Flakes in Gray Cast Iron**

   Fracture a piece of gray cast iron and compare the color of the surface with a machined surface.

The fractured surface appears gray like graphite because the fracture occurred almost entirely through the graphite flakes. The machined surface looks metallic because it contains only about 10% graphite.

3. **Austenitic Stainless Steel**

   A magnet provides a simple way of distinguishing austenitic stainless steels from the other grades. Austenitic stainless steel is not magnetic. Use a magnet on several pieces of stainless steel (i.e., dinnerware, a sink, hub caps, razor blades) and determine which are austenitic.

   If the stainless steel is not attracted to the magnet, then it is not austenitic stainless steel.

REFERENCES

7.1 J. Chipman, *Met. Trans.* Vol 3, 1972

7.2 *Making, Shaping and Treating Steels,* 9th ed., United States Steel Corp, 1971

7.3 *Atlas of Microstructures of Industrial Alloys,* Vol 7, *Metals Handbook,* 8th ed., American Society for Metals, 1972

7.4 W.F. Hosford, *Physical Metallurgy,* CRC, 2005

7.5 L.H. Van Vlack, *Elements of Materials Science,* 5th ed., Addison-Wesley, 1985

7.6 *Metallography, Structures and Phase Diagrams,* Vol 8, *Metals Handbook,* 8th ed., American Society for Metals, 1973

CHAPTER **8**

Ceramics

CERAMICS ARE NONMETALLIC, inorganic materials. Most ceramics can be grouped into several distinct groups. One group consists of crystalline compounds such as magnesium oxide, aluminum oxide, and boron nitride. Glasses form a second group. There also are materials formed by reactions involving water. Most ceramics are brittle: they fracture before they deform very much.

Crystalline Ceramics

Crystalline ceramics are compounds consisting of metal and nonmetal ions bonded either covalently or ionically. Most have high melting points and are very hard and brittle. Their tensile strengths are limited by brittle fracture but their compressive strengths are high. Because they retain high hardnesses at elevated temperatures, many are used for *refractories* (materials having a high heat resistance) such as crucibles and furnace linings. Oxidation at high temperatures is not a problem with refractory oxides as it is with refractory metals. Magnesia, alumina, and silica are used as refractories. (Oxides are named by substituting "a" for "ium" or "um" at the end of a metal's name. Thus, *alumina* is aluminum oxide, and *soda* is sodium oxide.) Ceramics also are used as tools for high-speed machining of metals. The high hardness of ceramics at room temperature leads to their use as abrasives, either as loose powder or bonded into grinding tools. The low ductility of ceramics limits the structural use of ceramics mainly to applications in which the loading is primarily compressive. Iron-containing ceramics are used as magnets.

The bonding strengths of ionic and covalent compounds depend on the valences of the metal and nonmetal. Compounds with higher valences (e.g., silicon carbide, silicon nitride) tend to be more strongly bonded than those of lower valences (e.g., sodium chloride, magnesium oxide); thus, they have higher melting points and higher hardnesses.

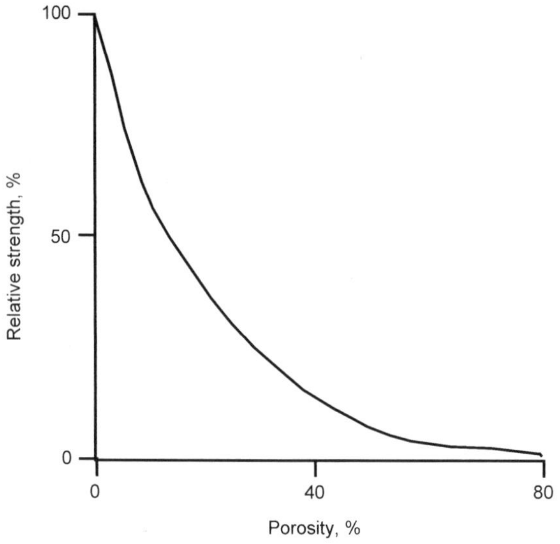

Fig. 8.1 Decrease of fracture strength with porosity. From data concerning stainless steel, iron, and plaster of paris. Source: Adapted from Ref 8.1

Most crystalline ceramics are prepared by pressing powder into the desired shape and sintering at a high temperature. The sintering causes the powder particles to coalesce, which results in some degree of porosity. Greater porosity causes less strength (Fig. 8.1).

Glasses

Glasses are amorphous materials, meaning they have no long-range order. However, glasses do have a great deal of short-range order. If crystallization is prevented during cooling, an amorphous glass will form. It will have short-range order inherited from the liquid. The critical cooling rate to prevent crystallization varies greatly from one material to another. Silicate glasses cannot crystallize unless the cooling rates are extremely slow. On the other hand, extremely rapid cooling is required to prevent crystallization of metallic glasses.

The basic structural units of silicate glasses are tetrahedra with Si^{+4} ions in the centers bonded covalently to O^{-2} ions at each corner. In pure silica all corner oxygen ions are shared by two tetrahedra (Fig. 8.2), resulting in a covalently bonded glass with a very high viscosity at elevated temperatures.

The compositions of typical commercial glasses are quite complex. Soda-lime glasses may contain 72% SiO_2, 14% Na_2O, 11% CaO, and 3% MgO. The Na^+, Ca^{+2}, and Mg^{+2} ions are bonded ionically to some of the corner O^{-2} ions (Fig. 8.3). With these alkali and alkaline earth oxides, not

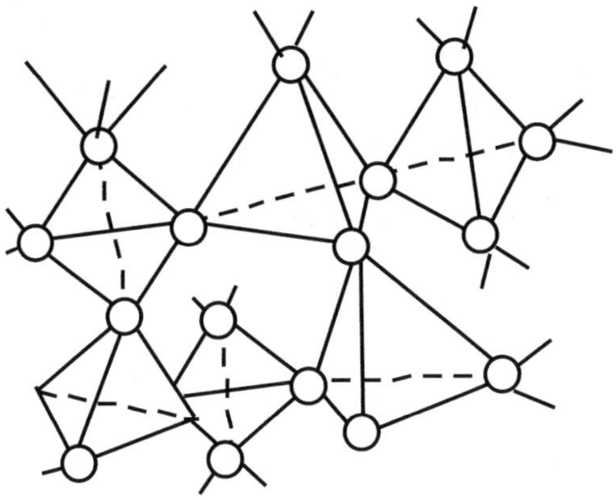

Fig. 8.2 Silica glass is composed of tetrahedra with four O^{-2} ions (shown by open circles) surrounding Si^{+4} ions. Each O^{-2} ion is shared by two tetrahedra. Source: Ref 8.2

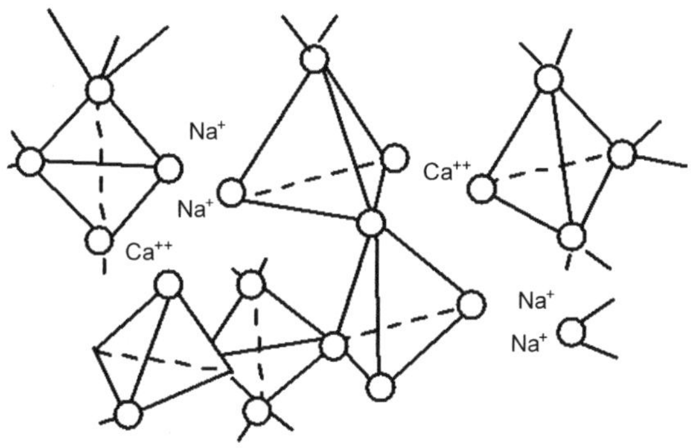

Fig. 8.3 Commercial glasses contain alkali and alkaline earth ions, which substitute ionic bonds for the covalent bonds between tetrahedra. Source: Ref 8.2

all of the oxygen ions are covalently bonded to two tetrahedra. This lowers the viscosity at high temperatures.

Glass Viscosity

The viscosities of glasses increase with the fraction of the number of oxygen ions that form bridges between two tetrahedra. Figure 8.4 is a plot of the viscosities of several glass compositions as a function of temperature. Above the glass transition temperatures, there is a strong temperature dependence of the viscosity.

Several temperatures are identified in terms of viscosity. In the *working range* (10^3 to 10^7 Pa·s), glass can be shaped economically. The *softening point* is defined by 4×10^6 Pa·s. Above this temperature a glass object will

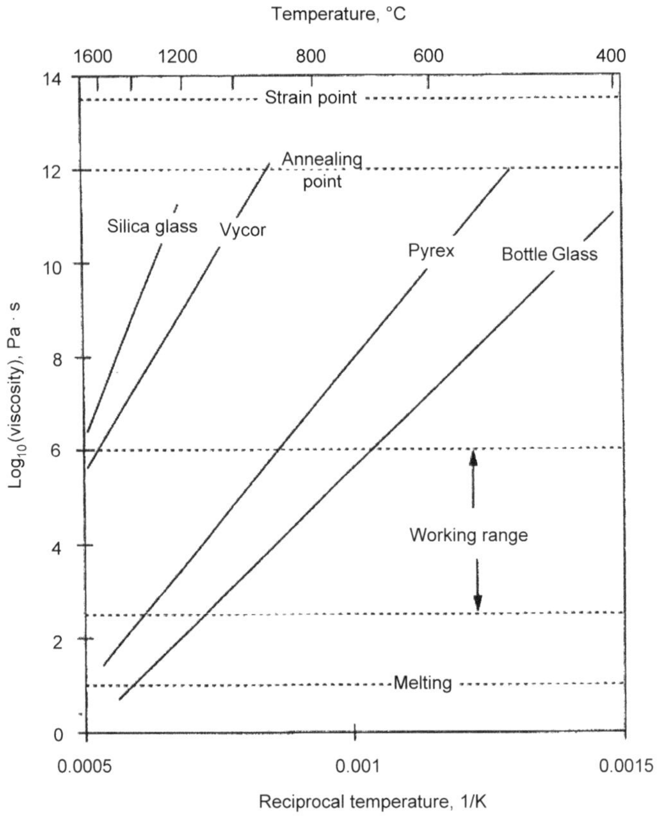

Fig. 8.4 Temperature dependence of viscosity for several glasses. The "working range" is the temperature range in which glasses can be economically shaped. The straight lines on the semi-log plot do not extend below the glass transition temperature.

creep under its own weight. Stress relief occurs in the *annealing range* ($10^{11.5}$ to $10^{12.5}$ Pa·s).

Below their glass transition temperatures, glasses should be regarded as solids rather than supercooled liquids even though they have the structure of a liquid. It is often said that the glass in the stained glass windows in European cathedrals is thicker at the bottom than at the top because they have flowed viscously over the centuries. Observable thickening at the bottoms by creep under self-weight would make the windows shorter. A 10% thickening at the bottom would be accompanied by approximately a 5% shortening, which would be much more obvious than the thickening. There would be a gap at the top of each pane, but there are none. The other problem with the story is that any creep would be much too small to produce observable deformation. The stress due to self-weight at the bottom of a pane, and the strain rate to produce 10% deformation over 750 years correspond to a viscosity less than could reasonably be expected by extrapolation to 20 °C (70 °F).

The true explanation is that the glass varied in thickness when it was installed. Until the 19th century, sheets of glass were made by spinning hot viscous globs on a rod. Centrifugal force from the spinning caused the glob to form into a disc (Fig. 8.5). The disc was thicker near the center than at the edges, so panes cut from it had a thickness variation. A skilled artisan naturally would install a pane with the thicker section at the bottom.

Thermally-Induced Stresses

Increased amounts of alkali and alkaline earth oxides in a glass increase its coefficient of thermal expansion as well as lower its viscosity. Table 8.1 shows coefficients of thermal expansion of several glass compositions, and Fig. 8.6 shows the relationship between the coefficient of thermal expansion and viscosity.

Fig. 8.5 Until the 19th century, panes of glass were made by spinning a rod with a glob of glass at the end and letting the centrifugal force form a disc from which panes could be cut. Source: Ref 8.3

Table 8.1 Compositions and coefficients of thermal expansion of several glasses

Glass	Composition, wt%(a)	$\alpha \times 10^{-6}$
Silica	100 SiO_2	0.5
Vycor	4 B_2O_3(b)	0.6
Pyrex	12 B_2O_3, 4 Na_2O, 4 Al_2O_3	2.7
Plate glass	13 CaO, 13 Na_2O, 2 MgO, 1 Al_2O_3	9.0

(a) In each case balance is SiO_2. (b) The composition of finished Vycor. The composition before forming has more of other oxides.

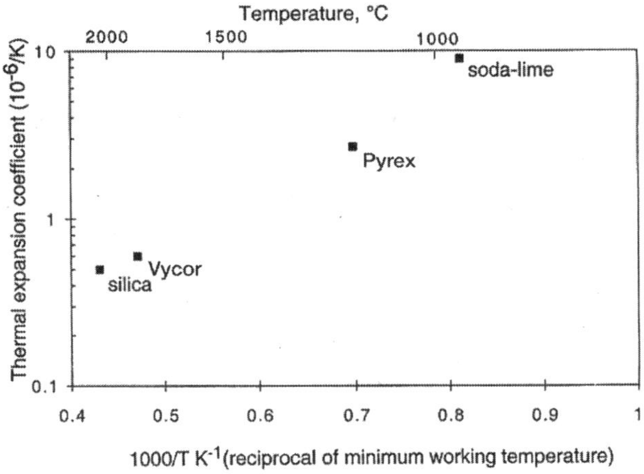

Fig. 8.6 The relation between thermal expansion coefficient and the temperature at which the viscosity is 10^7 Pa·s (10^8 poise). Compositions that promote lower working temperatures have higher coefficients of thermal expansion.

Glasses, like most ceramics, are susceptible to fracturing under stresses caused by temperature gradients. When there is a temperature change, two regions in close contact with each other must undergo the same dimensional change. The thermal expansion or contraction in one region must be matched by an elastic expansion or contraction in the other. The elastic dimensional change requires a stress, so internal stresses arise when there are different temperature changes in adjacent regions. If these stresses are high enough, fracture will occur.

Differences in resistance to thermal shock between different grades of glass are related to differences in their coefficients of thermal expansion. Larger stresses arise in materials with high coefficients of thermal expansion, so they are more susceptible to breaking under thermal shock. The thermal shock resistance of silica glass is considerably better than that of plate glass.

Silica-Rich Glass

A very high-silica glass product was developed to circumvent the difficulty in forming glasses that have a low coefficient of thermal expansion. The starting composition (62.7% SiO_2, 26.9% B_2O_3, 6.6% Na_2O, and 3.5% Al_2O_3) has a low enough viscosity at a reasonable temperature for the glass to be shaped. After the glass is shaped, it is heat treated between 500 and 750 °C (900 and 1380 °F). During this heat treatment a reaction separates the glass into two phases, one containing 96% SiO_2, 4% B_2O_3 and the other containing all of the Na_2O and Al_2O_3. The phase rich in soda and alumina is removed by acid etching, leaving a silica-rich glass with approximately

28% porosity. This can be used either as a filter or reheated to allow surface tension to produce a fully dense product.

Devitrification

If a glass is held for a long period at an elevated temperature, it may start to crystallize or *devitrify*. Devitrification of fused quartz (silica glass) to the crystalline silica is slow. Nucleation usually occurs at a free surface and often is stimulated by contamination from alkali ions such as sodium. With surface contamination, devitrification of fused quartz may occur at temperatures as low as 1000 °C (2000 °F). However, if the surface is clean, it rarely occurs below 1150 °C (2100 °F).

Some glasses may be intentionally crystallized to produce fine-grained ceramics. These materials are first formed as a glass into their final shapes at elevated temperatures, then heat treated at a lower temperature to allow nucleation of crystals, and finally heated to a somewhat higher temperature to allow growth of the crystals. Glass forming processes are used to obtain the final shape while producing a final product that is resistant to thermal shock because of a very low thermal expansion coefficient.

Photosensitive Glasses

Photosensitive glasses containing microcrystalline silver halides darken with exposure to ultraviolet light and lighten after the sunlight is removed. The darkening occurs when silver halides are reduced to metallic silver on exposure to ultraviolet light.

Other Inorganic Glasses

Borax, B_2O_3 (boron trioxide), forms a glass in which the basic structural elements are triangles with boron at the center surrounded by and covalently bonded to three oxygen atoms. Each of the oxygen atoms is shared by two triangles (Fig. 8.7).

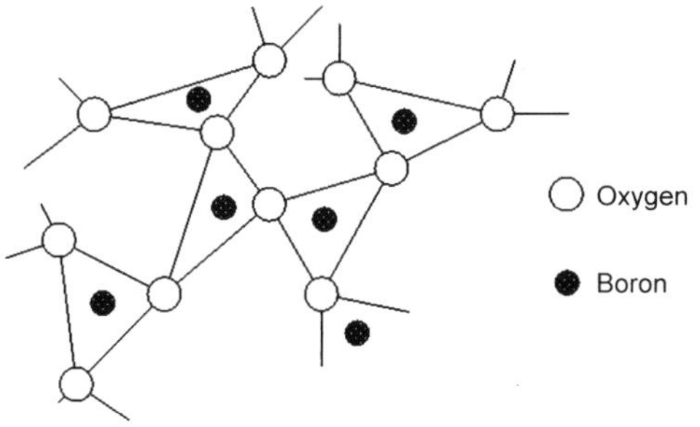

Fig. 8.7 Boron trioxide glass. Each boron atom is covalently bonded to three oxygen atoms, which form a triangle around the boron atom. Each oxygen atom is shared by two triangles. Source: Ref 8.2

There also are glasses consisting of long selenium (or tellurium) chains bonded with germanium or arsenic. In these glasses the basic structural units are chains that are cross-linked by germanium or arsenic. The structure of molten pure selenium and pure tellurium consists of long-chain molecules.

Residual Stresses

The fracture resistance of glass can be improved by inducing a pattern of residual stress with the surface under compression. The surface compression is balanced by a residual tension in the center. Residual compression in the surface increases fracture resistance because fractures almost always start at the surface, and because the stresses are highest at the surface if they involve any applied bending or torsion. Furthermore, defects are much more likely to be present at the surface. With residual compression at the surface, greater external loads can be tolerated before fracture occurs.

Compressive residual stresses can be induced in the surface either by cooling rapidly from high temperature or by chemical treatment. In the former process, called *tempering*, the glass is cooled by jets of air. During the cooling, the surface undergoes a thermal contraction before the interior. The different amounts of thermal contraction are accommodated by viscous flow of the interior. When the interior finally cools, dimensional compatibility is maintained by elastic contraction (compression) of the surface. Glass also may be *chemically tempered* by ion exchange. Glass is immersed in a molten salt bath containing potassium ions. Some of the K^+ ions diffuse into the glass, replacing Na^+ ions. Because the K^+ ions are larger than the Na^+ ions that they replace, the region affected is left in compression. One important difference between the two processes is the depth of the compressive layer. The depth of the region under compression is much less in the chemically tempered glass than in glass tempered by rapid cooling. The benefits of tempered glass are significant. Not only is tempered glass more resistant to fracture than untempered glass, but it also breaks into much smaller pieces, which are less dangerous than the large shards produced in fracture of untempered glass. Tempered glass is used for the side and rear windows of automobiles for these reasons. The windshields of automobiles are made from safety glass produced by laminating two pieces of glass with a polymeric material (polyvinyl butyral, PVB) that keeps the broken shards from causing injury. Figure 8.8 shows fracture patterns typical of untempered glass, laminated safety glass, and tempered glass.

Pottery

Pottery was the first man-made material. The oldest known pottery is dated between 27,000 and 23,000 B.C. Pottery is made by shaping wet

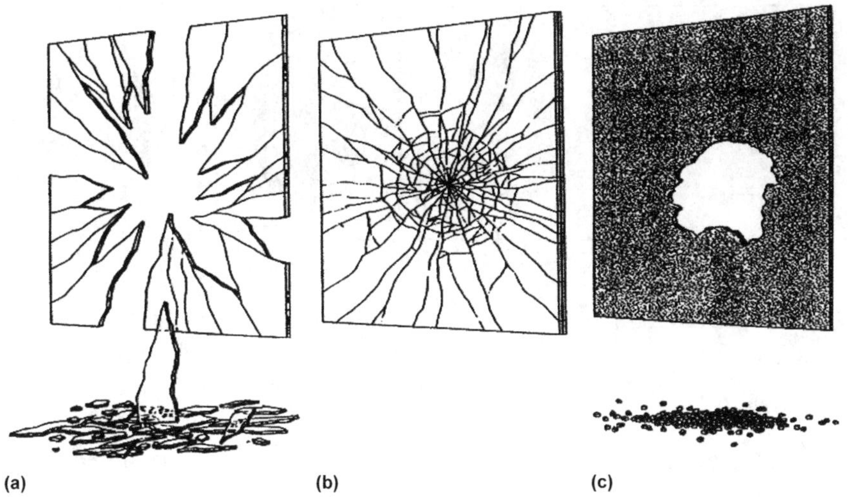

Fig. 8.8 Typical fracture patterns of three grades of glass. (a) Annealed (untempered) glass. (b) Laminated safety glass. (c) Tempered glass. Source: Ref 8.4

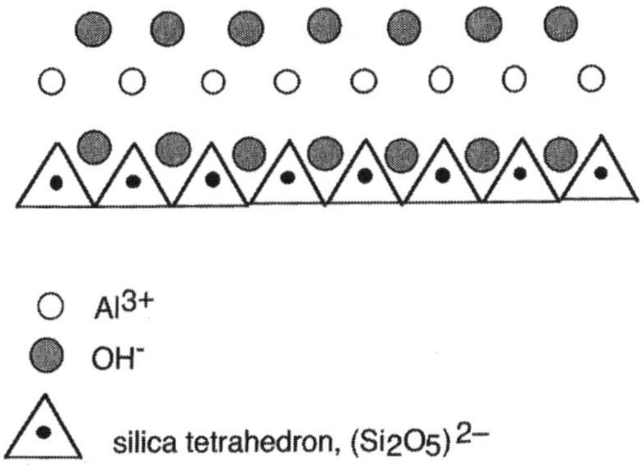

Fig. 8.9 Schematic of the molecular structure of clay. Source: Ref 8.2

clay, drying it, and then firing it at an elevated temperature that changes its chemical structure.

There are several forms of clay. All are aluminosilicates; many contain other elements. Figure 8.9 shows the platelike structure of *kaolinite,* $Al_2(Si_2O_5)(OH)_4$. Water absorbed between platelets allows them to slide easily over one another so clay can be formed easily when wet.

Typical clay mixtures used for ceramics often contain other materials including ground quartz (SiO_2) and a flux such as a feldspar, which is an aluminosilicate containing Na^+, K^+, and Ca^{+2}.

Processing of Clay Products

Clay can be shaped either by pressing into the desired shape or by *slip casting. Slip* is a suspension of clay in water made by adding a small

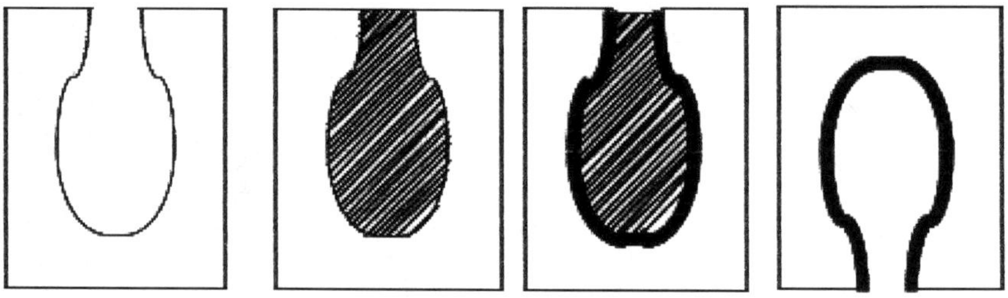

Fig. 8.10 Steps in slip casting. Source: Ref 8.2

amount of a *deflocculant* (often sodium silicate or soda ash), which allows the clay to be suspended in water. The slip is poured into a plaster-of-paris mold that absorbs water and causes the clay near the surface of the plaster to become solid. When this solid skin reaches the desired thickness, the mold is emptied and the resulting shape removed. Figure 8.10 illustrates the slip casting process for clay products.

The shaped clay is fired only after it is first air dried. Some shrinkage occurs during air drying. After air drying, the shape is said to be *green* and must be handled with care because it has little strength. Next it is fired, usually between 900 and 1400 °C (1700 and 2600 °F). Firing drives off the chemically combined water, with additional shrinkage but a great increase of strength. For pure kaolinite, the composition after firing is $3(Al_2O_3){\cdot}2(SiO_2)_2$, which is a mineral called mullite.

$$3Al_2(Si_2O_5)(OH)_4 \rightarrow 3(Al_2O_3){\cdot}2(SiO_2)_2 + 6H_2O$$

Figure 8.11 is the SiO_2-Al_2O_3 phase diagram.

Usually silica and feldspar are added to the kaolinite. During firing they form a silica-rich glass that bonds together the mullite crystals.

Earthenware, ovenware, stoneware, china, and porcelain are various forms of pottery that differ in composition and firing temperature. Earthenware basically is kaolinite with some silica and feldspar fired at lower temperatures than other clay products. It is relatively weak with porosity of 5 to 20%. It may be left unglazed as in flowerpots and soil pipe, or it may be glazed to make it water resistant. Stoneware is fired at higher temperatures, which reduces the porosity to less than 5%, giving it better strength. China is fired at even higher temperatures, which converts much of the silica, clay, and feldspar mixture into a glass. Increased firing temperatures lower the thermal expansion coefficient, α. Pottery with $\alpha < 4 \times 10^{-6}$/K can withstand air cooling after heating in an oven. To be able to withstand flame contact, the coefficient of thermal expansion should be less than 2×10^{-6}/K.

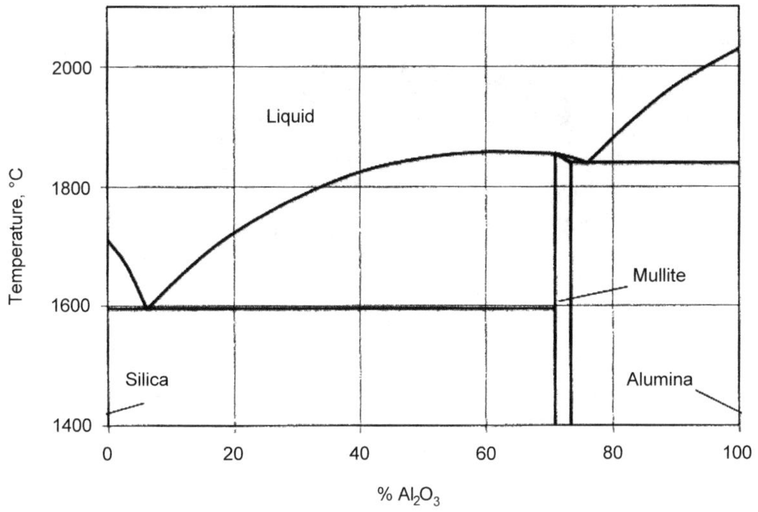

Fig. 8.11 The SiO_2-Al_2O_3 phase diagram showing mullite as an intermediate phase

Most pottery is glazed. After firing, a glaze is applied as a suspension of low-melting-point glass particles. Then the pottery is fired a second time to allow the glaze to melt and spread over the surface. Glazing has two functions: earthenware vessels would otherwise be unable to hold liquids due to porosity, and glazing also is done for decoration. To avoid cracking on cooling, the glaze should have a lower coefficient of thermal expansion than the pottery base so it will be left under a residual compressive stress. Sometimes, however, for aesthetic reasons glazes intentionally are formulated to produce surface cracks.

Hydration Reactions

Several materials react with water to form solids. Among these are plaster of paris, lime plaster, and Portland cement. Plaster of paris is the hemi-hydrate of calcium sulfate, $CaSO_4{\cdot}0.5H_2O$. It is obtained by heating gypsum, $CaSO_4{\cdot}4H_2O$, to approximately 150 °C (300 °F), which releases some of the hydrated water.

$$2CaSO_4{\cdot}4H_2O \rightarrow 2CaSO_4{\cdot}0.5H_2O + 7H_2O$$

When dry plaster powder is mixed with water to form a paste, it hydrates to form hard gypsum by the reverse reaction. The rate of the hydration reaction depends on the particle size of the plaster of paris because the reaction product surrounds the unreacted plaster of paris (Fig. 8.12). This slows the reaction because water must diffuse through the gypsum. Thicker gypsum layers slow the reaction; thus, hydration occurs more slowly with large particles.

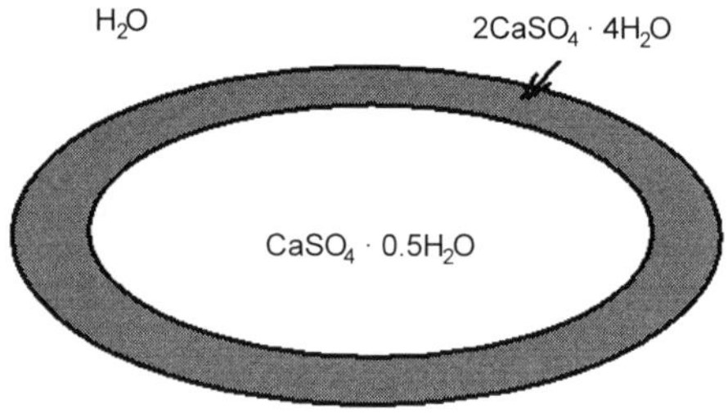

Fig. 8.12 In a typical hydration reaction, the product separates the reacting chemicals. Here water is separated from the plaster of paris by the gypsum formed. Source: Ref 8.2

Lime plaster is a mixture of calcium hydroxide and sand (or other inert fillers). Lime is made by heating limestone (calcium carbonate) to above 830 °C (1520 °F) to cause it to decompose into calcium oxide and carbon dioxide, $CaCO_3 \rightarrow CaO + CO_2$. When lime is hydrated with water, it forms lime plaster, $CaO + H_2O \rightarrow Ca(OH)_2$. With long periods of time, plaster reacts with atmospheric CO_2 to form calcium carbonate. Commercial plasters contain both plaster of paris and lime. Lime often is added to plaster of paris to control the rate of hardening. The rate of the hydration reaction depends on particle size as well as the fraction of each component.

Portland Cement

Portland cement contains several minerals, the two principal minerals being tricalcium silicate, $(CaO)_3 \cdot (SiO_2)$, which is abbreviated as C_3S, and dicalcium silicate $(CaO)_2 \cdot (SiO_2)$, abbreviated as C_2S. There also is some tricalcium aluminate, $(CaO)_3 \cdot (Al_2O_3)$, abbreviated by C_3A.

Dry powdered cement is mixed with water, which forms hydrates by the reactions:

$$Ca_3Al_2O_6 + 6H_2O \rightarrow Ca_3Al_2(OH)_{12}$$
$$Ca_2SiO_4 + xH_2O \rightarrow Ca_2SiO_4 \cdot (H_2O)_x$$
$$Ca_3SiO_5 + (x + 1)H_2O \rightarrow Ca_2SiO_4 \cdot (H_2O)_x + Ca(OH)_2$$

Cement also may contain some tricalcium aluminum ferrite, $(CaO)_3 \cdot (Al_2O_3) \cdot (FeO)$, abbreviated as C_3AF.

Hardening occurs by these hydration reactions in which water is chemically bound into hard compounds rather than by "drying." The rate of hardening is controlled by the composition of the mixture and by the particle size of the cement powder. Because heat is released during hydration, large structures must be cooled to slow the curing to prevent thermal gradients that would cause cracking.

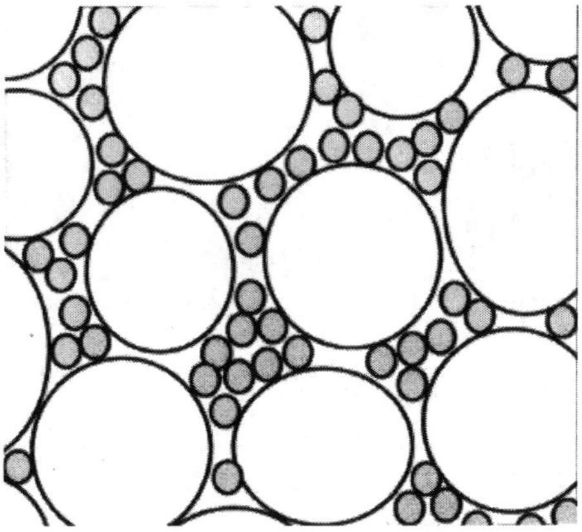

Fig. 8.13 Concrete consists of gravel particles (large open circles), sand (small gray circles), and cement paste filling the remaining space.
Source: Ref 8.2

Concrete Mixes

Concrete is a mixture of gravel, sand, and cement paste (Fig. 8.13). There are two reasons for adding gravel and sand to cement: sand and gravel particles are stronger than the hydrated cement, and they are significantly less expensive than cement. If the particles of gravel filled 60% of the space and sand filled 60% of the remaining space, the volume that the cement paste must fill would be only (100 – 60)(100 – 40) = 16%. However, because mixing is not complete, the ratio of sand to cement and the ratio of cement paste to sand should be higher.

A common concrete mix contains gravel, sand, cement, and water in the ratio of 3.1 gravel: 2.6 sand: 1.0 cement powder: 0.55 water by volume. With this mixture there is enough sand to fill the spaces between gravel particles and enough cement paste to fill the spaces between the sand.

Carbon

Carbon occurs in several forms including diamond, graphite, amorphous carbon, and fullerenes. None of these forms can be classified as a metal, ceramic, or polymer. Figure 8.14 shows the equilibrium between graphite, diamond, and liquid.

Diamond

Each carbon atom in diamond is covalently bonded to four other carbon atoms (Fig. 8.15). Very strong bonding makes diamond the hardest (10,000 kg/mm^2) known material. Diamond is used for cutting very hard materials. Diamond has an extremely high Young's modulus (1050 GPa,

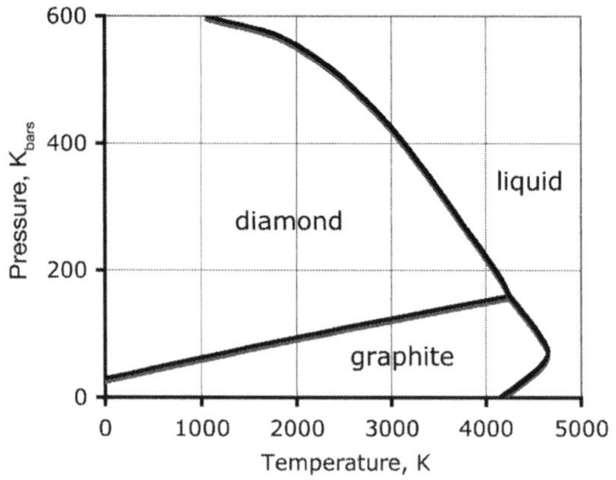

Fig. 8.14 Phase diagram showing the equilibrium phases of carbon. Source: Ref 8.2

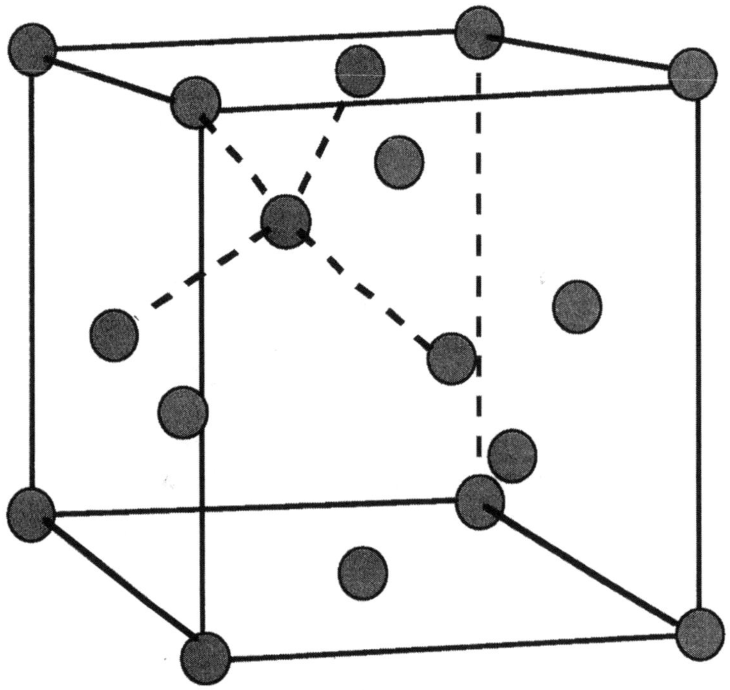

Fig. 8.15 The crystal structure of diamond. Each carbon atom is covalently bonded to four others.

or 15 × 10^4 ksi) and a very low coefficient of thermal expansion (1 × 10^{-6}/K). It has the highest thermal conductivity of all materials (2 kW/m·K compared with 401 W/m·K for copper), making it useful for dissipating heat. Its density (3.52 Mg/m^3) is considerably greater than that of graphite (2.25 g/m).

The first synthetic diamonds were made by subjecting carbon to very high pressures at high temperatures. Later it was learned that diamond also can be grown by chemical vapor deposition (CVD) under low pressure (1 to 27 kPa, or 0.15 to 3.92 psi.) Gasses including a carbon source and typically hydrogen are heated in a pressurized chamber and broken down, depositing diamond on exposed surfaces. Large areas (over 150 mm^2, or 0.23 $in.^2$) can be coated on a substrate. This allows CVD diamond films to be used as heat sinks in electronics and wear-resistant surfaces.

Graphite

The structure of graphite consists of sheets of carbon atoms arranged in a hexagonal pattern (Fig. 8.16). The bonding in the hexagonal sheets is like that in a benzene ring. Each carbon has two single bonds and one double bond. The sheets are bound to each other by weak van der Waals bonds. The ease with which sheets can slide over one another explains the lubricity of graphite. Because the double bond can move freely, the electrical conductivity in the plane of the sheet is very high, like that of a metal. The electrical and thermal conductivities perpendicular to the sheets are very low. The Young's modulus is very high in the planes of the sheet and very low perpendicular to them. The anisotropy of properties listed in Table 8.2 reflects the difference in bond strengths parallel and perpendicular to the sheets.

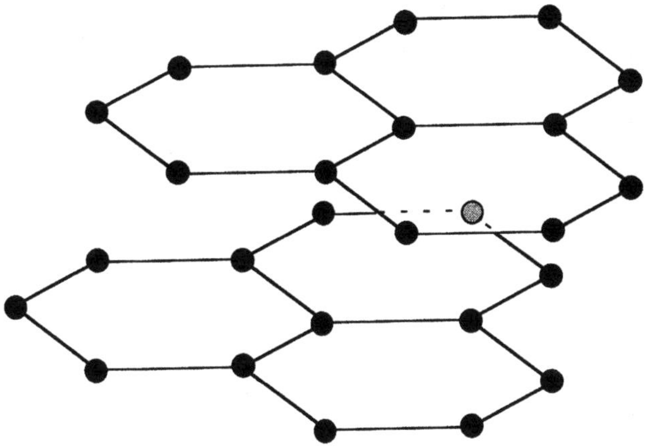

Fig. 8.16 The structure of graphite. Source: Ref 8.2

Table 8.2 Directional properties of graphite

Property	Perpendicular to c	Parallel to c
Electrical resistivity, ohm·m	2.5 to 5×10^{-6}	$3000 \times ^{-6}$
Thermal conductivity, W/m·K (Btu·in/(h·ft^2·°F)	398	2.2
Thermal expansion at 20 °C	(C^{-1}) 25×10^{-6}	Slightly negative $°C^{-1}$
Elastic modulus, GPa (ksi)	1060	35.5

Source: Ref 8.5

Carbon Fibers

Carbon fibers are thin graphite ribbons. They are made by pyrolizing polymeric precursors such as polyacrylonitrile (PAN), pitch, rayon, or other polymers that have carbon-carbon backbones. (Pyrolysis is the heating of carbonaceous materials in the absence of oxygen to drive off hydrogen.) Processing consists of several steps: stretching to align molecular chains; then heating to stabilize the orientation, and pyrolize and graphitize. The strengths and moduli are very high because they involve stretching carbon-carbon bonds. The properties of graphite fibers depend greatly on the nature of the precursor, its diameter, and the details of processing. Young's moduli of commercial carbon fibers vary from 200 to 700 GPa (29×10^3 to 102×10^3 ksi), and tensile strengths from 2 to 7 GPa (290 to 1015 ksi). Carbon fibers are used in epoxy-bonded composites.

Amorphous Carbon

Although soot and coal often are referred to as being amorphous, they actually contain small regions that are graphitic or diamond-like. Amorphous carbon can be produced by vapor sputter deposition. Some of the bonding is sp^2 (graphite-like) and some is sp^3 (diamond-like). (The "s" and "p" in sp^2 and sp^3 refer to sigma and pi bonds, respectively.) The ratio of the two types of bonds may vary considerably. Amorphous carbon contains a high concentration of dangling bonds, causing deviations of the interatomic spacing of more than 5%, and noticeable variations in bond angles.

Fullerenes

Until 1985, the only known elemental forms of carbon were diamond, graphite, and amorphous carbon. Then in 1985, Kroto et al. (Ref 8.6) announced the discovery of C_{60}, a spherical arrangement of carbon atoms in hexagons and pentagons (Fig. 8.17). They called this form the *Buckminsterfullerene* after the architect, Buckminster Fuller, who developed the geodesic dome. The name for this type of carbon molecule has since been shortened to *fullerene,* and the molecules are commonly called *Buckyballs.* Since their discovery, it has been found that fullerenes can be made in quantity from electrical arcs between graphite electrodes. Approximately 75% of the fullerenes produced by arcs are C_{60}, 23% are C_{80}, with the rest being even larger molecules. Approximately 9000 fullerene compounds are known.

Nanotubes

Carbon nanotubes are in the family of fullerenes. They can be regarded as hexagonal sheets of carbon atoms (*graphene* sheet), rolled up to make a cylinder and capped at the ends by one-half of a Buckyball (Fig. 8.18).

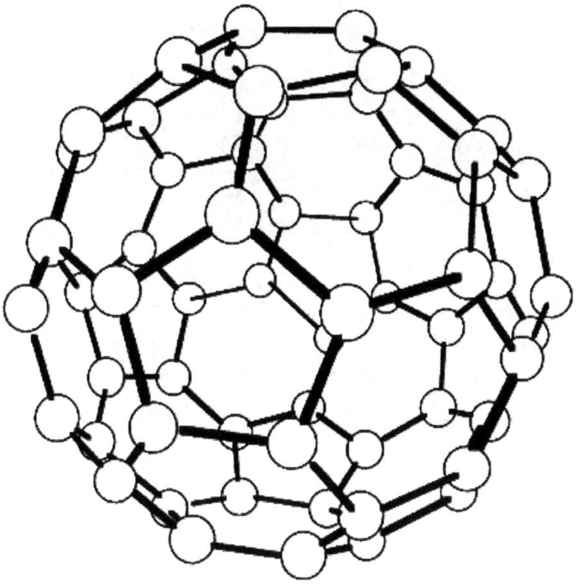

Fig. 8.17 A C_{60} Buckyball, which has 60 carbon atoms arranged in hexagons and pentagons. The arrangement is the same as that on a soccer ball. Source: Ref 8.2

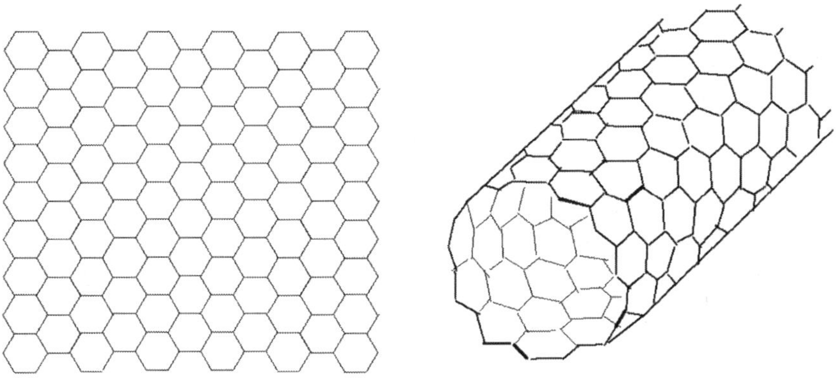

Fig. 8.18 Single-wall nanotubes can be thought of as being made from rolled-up chicken wire. Source: Ref 8.2

Tubes typically have diameters of approximately 1 nm (0.039 μin.). The diameter of the smallest nanotube corresponds to the diameter of the smallest Buckyball (C_{60}). The length-to-diameter ratio typically is approximately 10^4.

Nanotubes can be made by arc discharge, laser ablation, or chemical vapor deposition (CVP). There are a number of potential uses of fullerenes, including using nanotubes for field effect transistors. Nanotubes decorated with metal atoms have a great potential for hydrogen storage in fuel cells.

NOTES OF INTEREST

- Photosensitive glass was first developed by Donald Stookey at Corning Glass in 1937. A U.S. patent was issued in 1950.
- Portland cement was developed in England. Its name comes from its similarity to the stones on the Isle of Portland. A British patent on its composition was issued in 1824.

In another potential use of fullerenes, A_3C_{60} compounds in which A is an alkali (potassium, rubidium, cesium, sodium) are superconductors. Lastly, sieves made of fullerenes have been suggested that allow biological compounds to pass through but do not permit entrance of larger viruses.

Exercise

1. **Changes during Processing of Clay**

   Make an object out of clay. Weigh it. Weigh it again after a day. Then heat it to 400 °C (750 °F) for one hour and let it cool. Weigh it again. Explain the differences in weight.

   The weight decreases during the day after forming because of the evaporation of water. The weight decreases further when heated to 400 °C as more water is driven off. Hydration levels decrease both times.

REFERENCES

8.1 R.L. Coble and W.D. Kingery, *J. Am. Ceram. Soc.,* Vol 29, 1956

8.2 W.F. Hosford, *Materials for Engineers,* Cambridge U. Press, 2008

8.3 *Diderot Pictorial Encyclopedia of Trades and Industries,* Plate 245, 1758

8.4 *Engineering Materials Handbook Volume 4: Ceramics and Glasses,* ASM International, 1991

8.5 H.O. Pierson, *Handbook of Carbon, Graphite, Diamond and Fullerenes,* Knovel, 2001

8.6 H. Kroto, J. Heath, S. O'Brien, R. Curl, and R. Smalley, *Nature,* Vol 318, 1985

Elementary Materials Science
William F. Hosford

CHAPTER 9

Polymers

THE WORD *polymer* is the scientific name for what is commonly called a plastic. The word *polymer* comes from the Greek *poly* (meaning many) and *mer* (meaning parts). Polymers are very large molecules with many repeating parts, or mers. For example, polyethylene is composed of many repeating ethylene *mers*. It is formed by the reaction:

$$nCH_2 = CH_2 \rightarrow |\text{-}CH_2\text{-}CH_2\text{-}|_n$$

where *n* is a very large number. This reaction is called an *addition reaction*. It must be initiated by breaking the double carbon-carbon bond in the ethylene monomer. This may be broken by stimulation of an initiator such as a radical, R^-, which produces another radical that can stimulate yet another ethylene molecule in the reaction:

$$R^- + CH_2{+}CH_2 \rightarrow R^-(+\ CH_2\text{-}CH_2)^-$$

Another type of polymerization reaction involves two different monomers and produces a small byproduct. An example is the reaction between adipic acid, H–O–C=O–$(CH_2)_4$–O+C–O–H, and hexamethylenediamine, NH_2–$(CH_2)_6NH_2$:

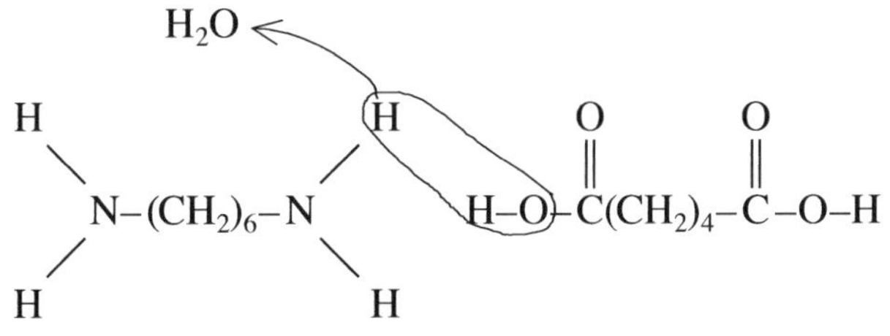

that produces nylon with a water molecule as a byproduct. This is repeated many times to make a long molecule. This type of reaction is called a *condensation reaction* because in the laboratory the resulting water condenses in a cool part of the apparatus.

Thermoplastics

The two examples above produce linear molecules. Such polymers are said to be *thermoplastic* because they soften on heating. Table 9.1 gives a number of examples of linear thermoplastics.

Other thermoplastics have more complex backbones. Among these are nylon, polyester (PET), polycarbonate, cellulose, and poly paraphenylene terphthalamide (PPTA). Figure 9.1 illustrates the molecular structure of these.

Silicones are polymers with -Si-O-Si-O- backbones.

Elastomers are rubberlike polymers. They have carbon backbones that include carbon-carbon double bonds. Figure 9.2 shows the repeating unit. The radicals, R, are listed in Table 9.2.

In all cases there is weak van der Waals bonding between neighboring chains. This allows heating to melt thermoplastics.

Thermosetting Polymers

In contrast to thermoplastics, some polymers form three-dimensional structures that cannot be melted, and where all mers are tied together by covalent bonds. Such polymers are said to be *thermosetting*. Once formed, their shape cannot be changed. Figure 9.3 is a schematic sketch of such a network. The open circles represent monomers that are tri-functional. Some have reacted with three smaller monomers (black) while others have reacted with only two.

Table 9.1 Polymers of the general form

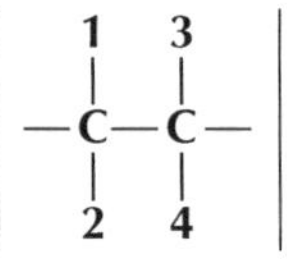

Polymer name	1	2	3	4	Abbreviation
Polyethylene	H	H	H	H	PE
Polyvinyl chloride	H	H	H	Cl	PVC
Polyvinyl alcohol	H	H	H	OH	PVA
Polypropylene	H	H	H	CH_3	PP
Polyvinyl acetate	H	H	H	$O\!-\!\overset{\overset{O}{\|}}{C}\!-\!C\!-\!H_3$	PVAc
Polystyrene	H	H	H		PS
Polyacrylonitrile	H	H	H	N	PAN
Polyvinylidene	H	H	Cl	Cl	PVDC
Polyethylene methacrylate	H	H	CH3	...	PMMA
Polyisobutylene	H	H	CH3	CH_3	PIB
Polytetrafluoroethylene	F	F	F	F	PTFE

 symbolizes a benzene ring.

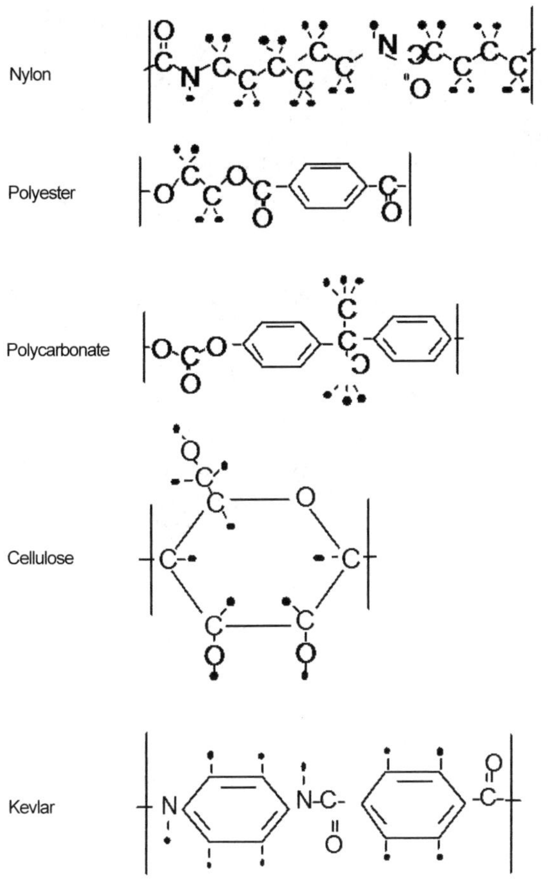

Fig. 9.1 Structure of several linear polymers. Kevlar is a registered tradename of E.I. du Pont de Nemours and Company. Source: Ref 9.1

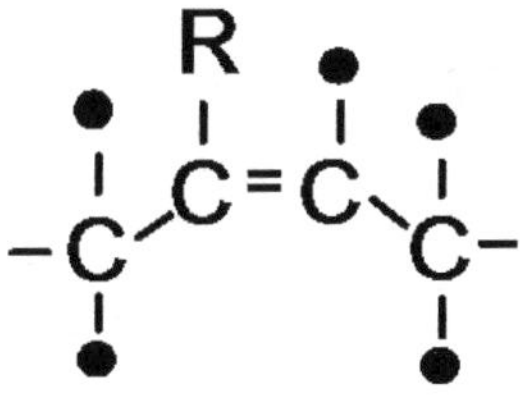

Fig. 9.2 Basic structure of rubbers. The filled circles are hydrogen atoms and *R* stands for a radical. Source: Ref 9.1

Table 9.2 Several rubbers

R	Rubber
H	Polybutadiene (buta-rubber)
Cl	Polychloroprene
CH_3	Polyisoprene (natural rubber)

Among common thermosetting polymers are phenol formaldehyde (Fig. 9.4) and urea formaldehyde (Fig. 9.5). Polyesters and epoxies also are thermosetting.

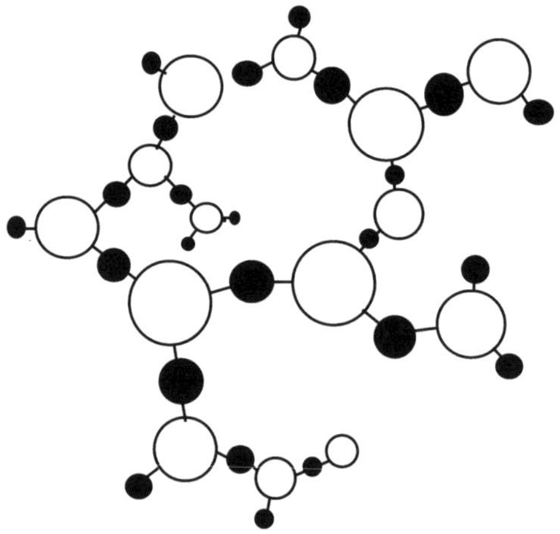

Fig. 9.3 Schematic illustration of a three-dimensional network formed by two types of monomers. The open circles represent a large monomer, and the black circles another smaller monomer. Source: Ref 9.1

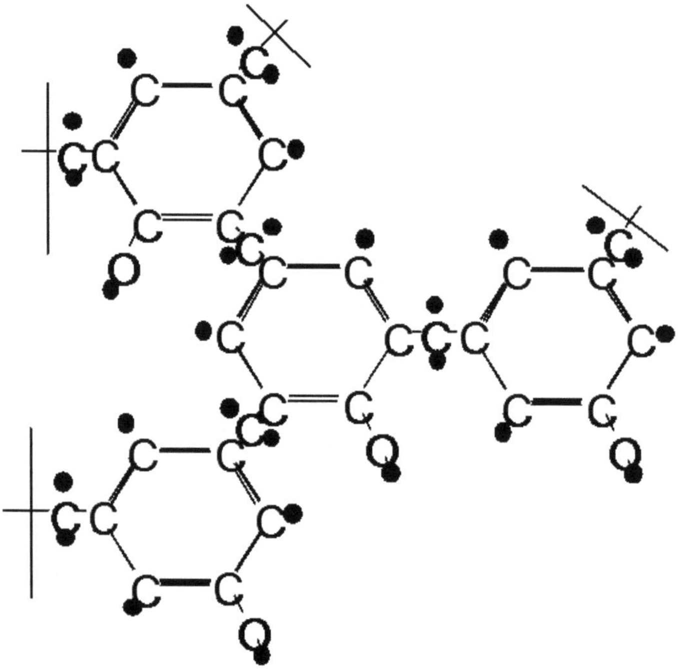

Fig. 9.4 The structure of phenol formaldehyde. Source: Ref 9.1

Degree of Polymerization and Molecular Weight

The *degree of polymerization* is the average number of mer units per molecule. It is the molecular weight of the polymer divided by the molecular weight of the mer.

Both strength and melting temperature increase with molecular weight. Figure 9.6 shows this effect for the paraffin series, C_xH_{2x+2}.

Branching

Linear polymers may form branches, as illustrated in Fig. 9.7.

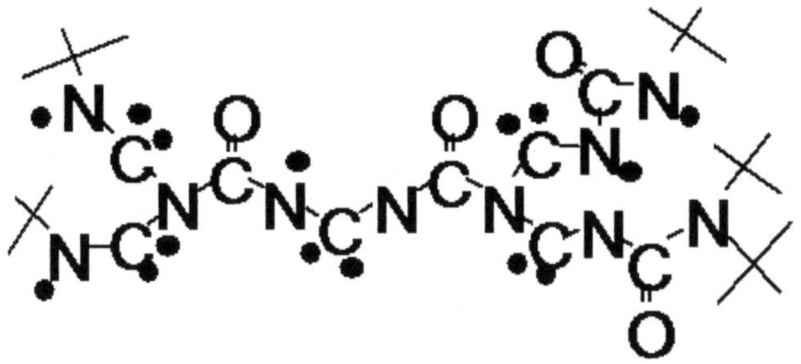

Fig. 9.5 The structure of urea formaldehyde. Black dots are hydrogen atoms. Source: Ref 9.1

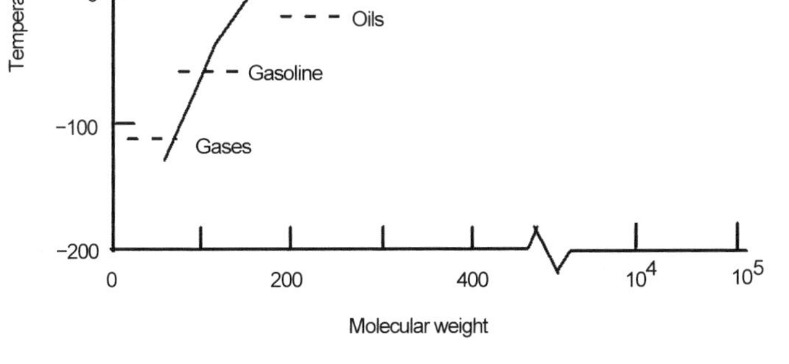

Fig. 9.6 Melting point increase of paraffin series with molecular weight. Source: Ref 9.1

Cross-linking and Stereoisomerism

Cross-links may be formed between molecular chains. Rubber is vulcanized with sulfur to form cross-links between polyisoprene molecules (Fig. 9.8). The hardness of the rubber increases with the number of cross-links.

Sometimes there is more than one distinguishable arrangement of side groups around the carbon-carbon chains in molecules of the same chemical formula. Rotation around the carbon-carbon bond cannot change these distinguishable arrangements. One example is polypropylene. If all of the CH_3 groups are on the same side of the chain, the polymer is said to be *isotactic*, whereas if they are on opposite sides, the polymer is said to be *syndiotactic* (Fig. 9.9). If the position of the CH_3 groups is random, the polymer is *atactic*.

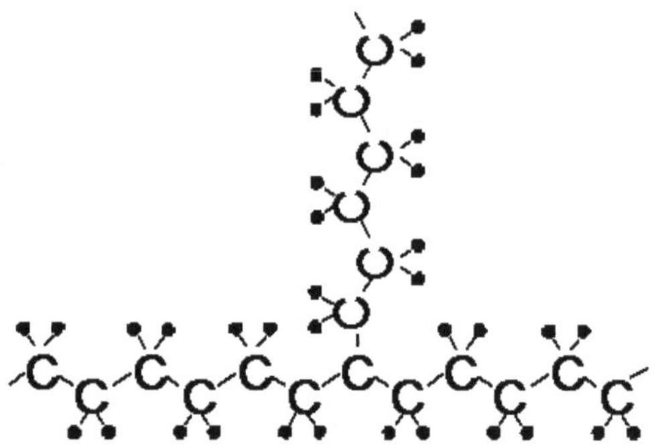

Fig. 9.7 Branched polyethylene. Source: Ref 9.1

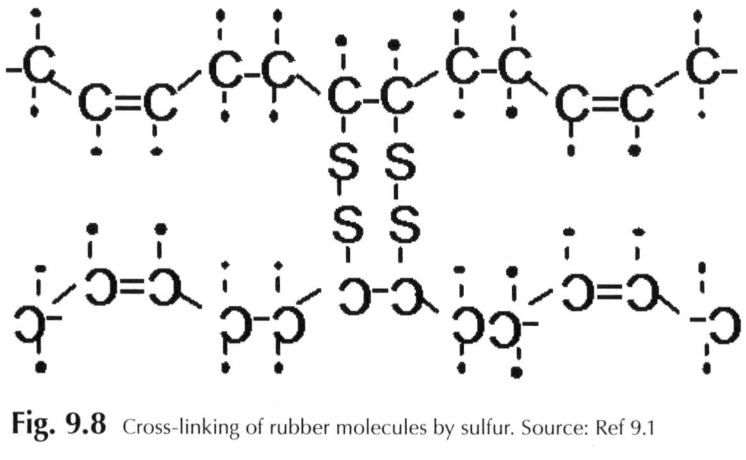

Fig. 9.8 Cross-linking of rubber molecules by sulfur. Source: Ref 9.1

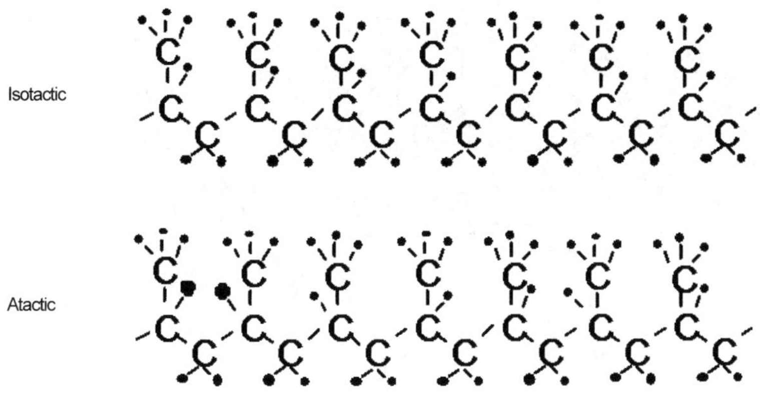

Fig. 9.9 Tacticity of linear polymers. Ref 9.1

Copolymers

Many commercial plastics are copolymers, or polymers in which two or more monomers are polymerized together. Acrylonitrile butadiene styrene (ABS) is a copolymer of acrylonitrile, butadiene, and styrene that incorporates some of the rubberlike properties of buta-rubber. What is commercially called "vinyl" is a copolymer of polyvinyl acetate and polyvinyl chloride (PVC). The properties of vinyls depend on the proportions of vinyl acetate and vinyl chloride. Pure PVC is rigid and brittle. Its use is limited to pipes and siding. With increasing amounts of vinyl acetate, it is used for insulating wires, molded parts, and as a lacquer.

Molecular Configuration

Linear polymers may be either amorphous or crystalline, or may contain both amorphous and crystalline regions. One may think of the molecules in amorphous regions of a polymer as being like cooked spaghetti. In crystalline regions, the molecules are parallel and fit closely together. Figure 9.10 illustrates these two possibilities.

The crystalline regions have a higher density and are stiffer than the amorphous regions. The factors that tend to prevent crystallinity are those that decrease the order of the polymer chains such as side branching, cross-linking, large side groups, irregular spacing of side groups (being atactic), and copolymerization. As a linear polymer is stretched, the molecules tend to become aligned with the direction of stretching. This increases the crystallinity and stiffness. Some polymers are highly crystalline but others have little or no crystallinity. Branching, large side groups, and copolymerization all tend to suppress crystallinity. There is no branching in high-density polyethylene, but low-density polyethylene is highly branched.

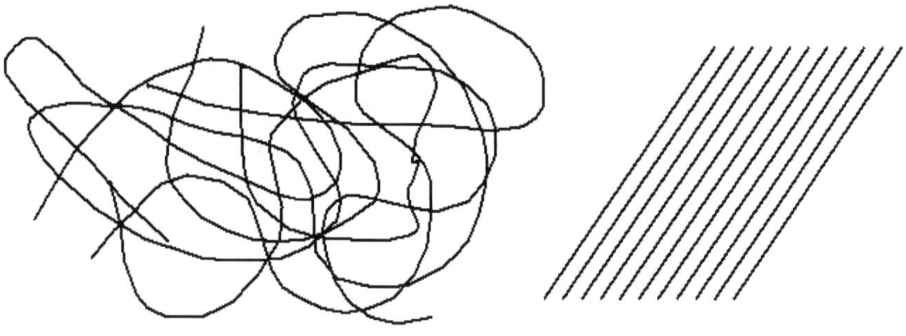

Fig. 9.10 In amorphous regions (left) the molecules are randomly oriented, while in crystalline regions (right) the molecules are aligned. Source: Ref 9.1

Glass Transition

The bonds between neighboring molecular chains are weak van der Waals bonds. Therefore, most polymers do not readily crystallize. Instead, on cooling they form an amorphous glass at a glass transition temperature. The relative positions of molecules are frozen so the amorphous glass is not a liquid. The change of specific volume on cooling is schematically illustrated in Fig. 9.11.

The glass transition temperatures of several polymers are given in Table 9.3.

At temperatures below their glass transition temperature, polymers tend to have a high Young's modulus and be brittle. Figure 9.12 shows the sharp drop of Young's modulus for PVC near its glass transition temperature. Clearly the glass transition temperature is near 80 °C (180 °F) and depends slightly on the time of loading.

The reason that the modulus is time dependent is illustrated by Fig. 9.13. For the same stress, the strain increases with the time of loading so the modulus drops.

Additives

Because the carbon-carbon bonds of polymers can be broken by ultraviolet radiation; carbon black, titanium dioxide, and other pigments often are added to block sunlight. Dyes do not perform this function. Other examples of additives are fillers such as wood flour and silica, which often are used to decrease price; and hard particles such as silica, which increase wear resistance. Lastly, small molecules called *plasticizers* are added to polymers to decrease the glass transition temperature and elastic modulus.

In the past, halogen-containing compounds were used for flame retardants, but because of toxicity and environmental concerns these compounds are being replaced by phosphorous-, nitrogen-, and silicon-containing organic compounds, inorganic compounds, and nanomaterials.

Degradation

There is a spectrum of interactions between plastics and solvents. There is no interaction of polyethylene or polyethylene terephthalate with water.

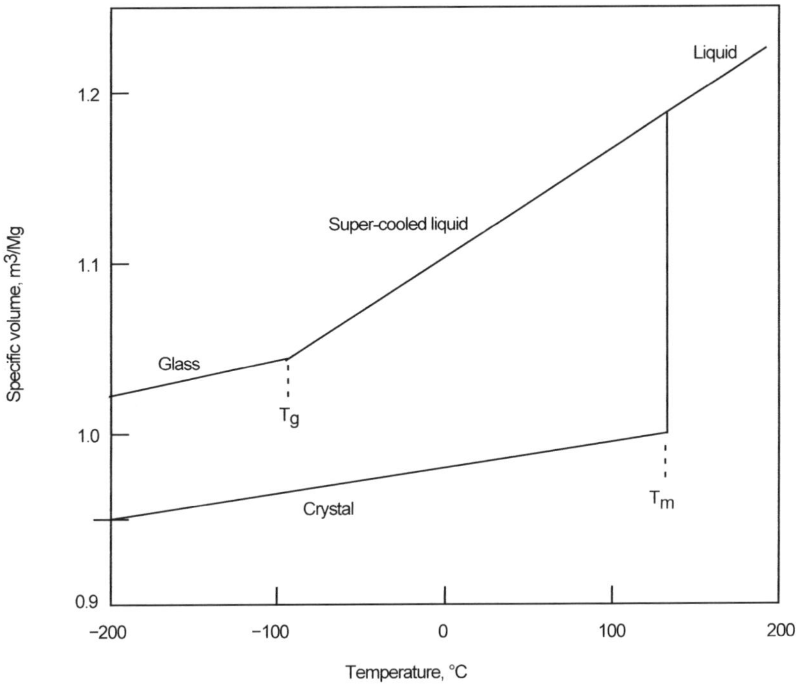

Fig. 9.11 Change of the specific volume of polyethylene with temperature. If it does not crystallize at the melting temperature, polyethylene will remain a supercooled liquid until it reaches its glass transition temperature. Source: Ref 9.1

Table 9.3 Glass transition (T_g) and melting (T_m) temperatures of several polymers

Polymer	T_g, °C (°F)	T_m, °C (°F)
Polyethylene (high density)	−120 (−184)	140 (284)
Polybutadiene	−70, (−94)	(a)
Polypropylene	−15 (5)	175 (347)
Nylon 6/6	50 (122)	265 (59)
Polyvinyl chloride	85 (185)	210 (410)(a)
Polystyrene	100 (212)	240 (464)(a)

(a) Difficult to crystallize

Other plastics will absorb a solvent and swell. Examples include nylon in water and PVC in ketones. Some solvents may dissolve certain polymers. Polyvinyl alcohol dissolves in water. Polyvinyl chloride pipes are joined together with solvents that evaporate to form welds. Gasoline will cause hydrocarbon rubbers to swell. Alcohol attacks polyurethanes.

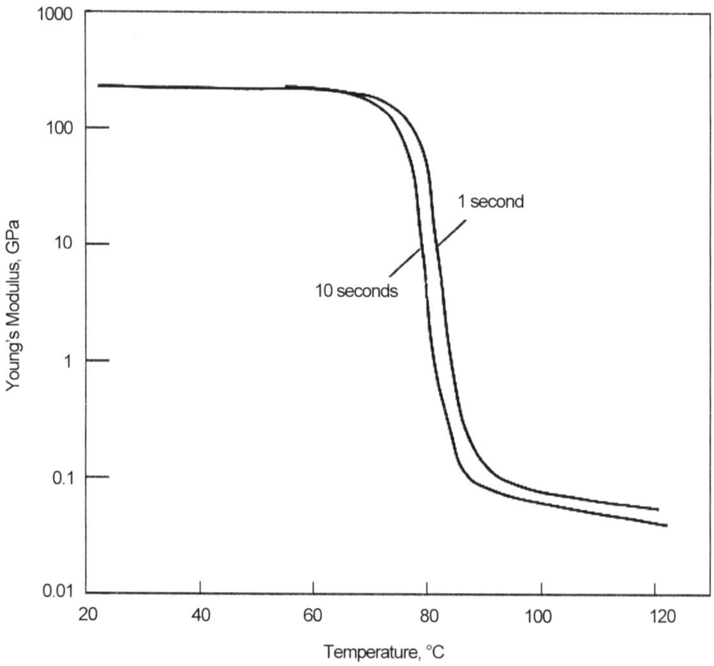

Fig. 9.12 Young's modulus of polyvinyl chloride (PVC) is approximately three orders of magnitude below the glass transition temperature rather than above it. It depends only slightly on the rate of loading. Source: Ref 9.1

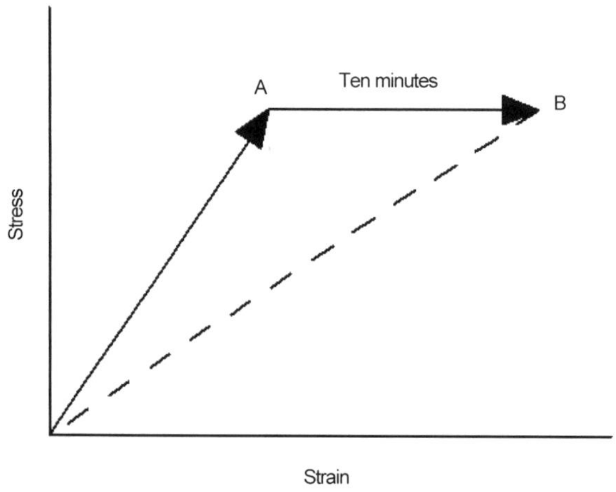

Fig. 9.13 At constant stress, the strain increases so the modulus, σ/ε, decreases with time of loading. Source: Ref 9.2

The double bonds of vulcanized rubber are susceptible to attack by ozone. The *scission reaction* involves reaction with ozone:

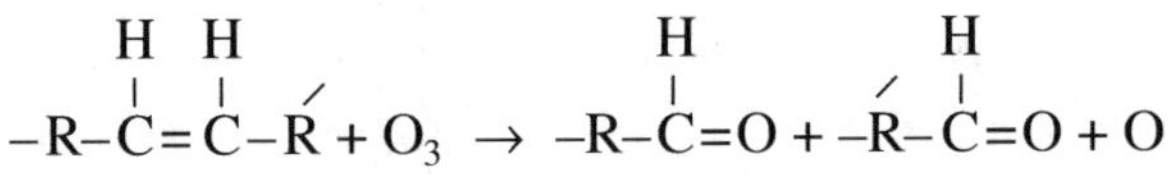

A combination of stress and environments containing some chemicals may lead to stress cracking.

Properties and Uses of Polymers

Polyethylene Terephthalate (PET)—Often Simply Called 'Polyester.' The density of PET is 1.35 to 1.38 g/cm^3 (84.27 to 86.15 lb/ft^3), melting temperature (T_m) = 260 °C (500 °F), and glass transition (T_g) = 75 °C (167 °F). It is tough, provides a barrier to water and gases, and finds uses as beverage bottles and food containers, yarns and polyfill, and film that is used for its dielectric properties.

High-Density Polyethylene (HDPE). The density of HDPE is 0.94 to 0.96 g/cm^3 (58.68 to 59.93 lb/ft^3), T_m = 130 °C (270 °F), and T_g = –90 °C (–130 °F). It is highly crystalline, giving it a high modulus of elasticity and making it translucent. It is tough, strong, easy to process, and resistant to water and chemical attack. It finds uses as bottles for milk, bleach, and motor oil.

Low-density polyethylene (LDPE) contains many branches that inhibit crystallization. Its density is 0.92 to 0.94 g/cm^3 (57.43 to 58.68 lb/ft^3), T_m = 170 °C (340 °F), and T_g = –110 °C (–170 °F). It is tough, flexible, and transparent, and finds wide usage in wire insulation, squeezable bottles, furniture, and carpets.

Polyvinyl Chloride (PVC)—Often Simply Called 'Vinyl.' The density of PVC is 1.32 to 1.42 g/cm^3 (82.40 to 88.64 lb/ft^3), T_m = 212 °C (414 °F), and T_g = 87 °C (189 °F). In unplasticized forms it is hard and brittle. It is used for building materials such as siding, window frames, and piping systems. As plasticized and copolymerized with polyvinyl acetate, it is used for clothing, upholstery, flexible tubing, and nonfood containers. Its flexibility depends on how much polyvinyl acetate has been copolymerized with it. It has stable electrical properties, is easy to blend and process, and is suitable for injection molding. Polyvinyl chloride is flame resistant but the chlorine in it produces HCl when PVC is burned.

Polystyrene (PS)—Contains a Benzene Ring. The density of PS is 1.05 g/cm^3 (65.54 lb/ft^3), T_m = 240 °C (460 °F), and T_g = 95 °C (203 °F). It is brittle and hard. It is molded into rigid objects, but the greatest use is as foam for insulation in construction and cups, and for packaging ("peanuts" and rigid shapes).

Polypropylene (PP). The density of PP varies from 0.85 g/cm^3 (53.06 lb/ft^3) for amorphous PP to 0.95 g/cm^3 (59.30 lb/ft^3) for crystalline PP; T_m = 165 °C (329 °F); and T_g = –10 °C (14 °F). Its resistance to chemical attack makes it suitable for linings that must be resistant to chemicals, sinks, ducts, barrels, tanks, pump components, and prosthetic devices. As a fiber it is used in "polypro" clothing.

Polymethyl Methacrylate (PMMA). The density of PMMA is 1.19 g/cm^3 (74.28 lb/ft^3), T_m = 130 to 140 °C (270 to 280 °F), and T_g = 105 °C (221 °F). It is highly transparent. It has a higher impact strength than glass but is softer and is used wherever toughness and transparency are required. Applications include shields for hockey, aircraft windows, colored glasses, and eyeglass lenses.

Nylon 6/6. There are a number of nylon compounds, which vary in the length of the monomers. The most common is nylon 6/6, which is made from two monomers, hexamethylenediamine and adipic acid, each of which contains six carbon atoms. The density of nylon 6/6 is 1.13 to 1.15 g/cm^3 (70.54 to 71.79 lb/ft^3), T_m = 265 °C (509 °F), and T_g = 57 °C (135 °F). It is used as a fiber in clothing and rope, as well as molded into gears and other objects that require strength and low friction.

Polycarbonate. The density of polycarbonate is 1.20 to 1.22 g/cm^3 (74.91 to 76.16 lb/ft^3), T_m = 267 °C (513 °F), and T_g = 150 °C (300 °F). It is very tough and transparent. It is used in applications requiring toughness and heat resistance such as bulletproof glass, injected molded and extruded shapes, film, signs, and football helmets.

Polyvinylidene Chloride. The density of polyvinylidene chloride is 1.7 g/cm^3 (106.12 lb/ft^3), T_m = 169 °C (336 °F), and T_g = –17 °C (1.4 °F). Its principal use is as plastic food wrap.

Aramids. There are three common grades of aramids. The density of aramids is 1.44 g/cm^3 (89.89 lb/ft^3), T_m = 640 °C (1180 °F), and T_g = 375 °C (707 °F).

Aramids are an aromatic polyamide. One of the grades typically is used for reinforcements in tires and other rubber mechanical goods. Another of the grades of aramids is used in industrial applications such as cables, replacement for asbestos, brake linings, and body armor. A third grade, considered to have the greatest tensile strength of all the aramids, is used in applications such as plastic reinforcement for boat hulls, airplanes, bicycles, bulletproof vests, trampolines, and tennis rackets.

Polytetrafluoroethylene (PTFE). The density of PTFE is 2.2 g/cm^3 (137.34 lb/ft^3), T_m = 327 °C (621 °F), and T_g = –110 °C (–166 °F). It decomposes at 350 °C (662 °F), so it cannot be molded. Because of its extremely low coefficient of friction it is used for coating armor-piercing bullets, bushings, bearings, and frying pans. A breathable PTFE fabric variant with micropores is used for raingear.

Phenol Formaldehyde. Although its use has declined because of cost, phenol formaldehyde is still used for small parts such as disc cylinders, electric plugs, and switches.

Urea formaldehyde is commonly used for electrical appliance casings such as desk lamps. It largely has replaced phenol formaldehyde.

Polyurethanes are used as both flexible and rigid foams. Flexible foam, often referred to as "foam rubber," is used for auto seats, chairs, and mattresses; rigid foam is used in surfboards. Solid polyurethanes are used for roller blade wheels, tables, and furniture, and for encasing electronic components. Polyurethanes also are the basis of some varnishes.

Acrylonitrile butadiene styrene (ABS) combines the strength and stiffness of acrylonitrile and styrene with the flexibility of butadiene rubber. It is used for keyboards. Because of its toughness and impact resistance, it is used in canoe hulls.

Epoxies. An epoxy is a thermosetting resin that polymerizes when mixed with a catalyst or hardener. Most common resins are produced from a reaction between epichlorohydrin and bisphenol-A. Epoxy materials tend to harden somewhat more gradually than polyesters. Epoxies are widely used as adhesives.

Silicones are analogous to linear polymers. They have a Si-O backbone with organic radicals or hydrogen attached to the silicon atoms (Fig. 9.14). Silicones are stable up to 250 °C (480 °F) and are resistant to sunlight,

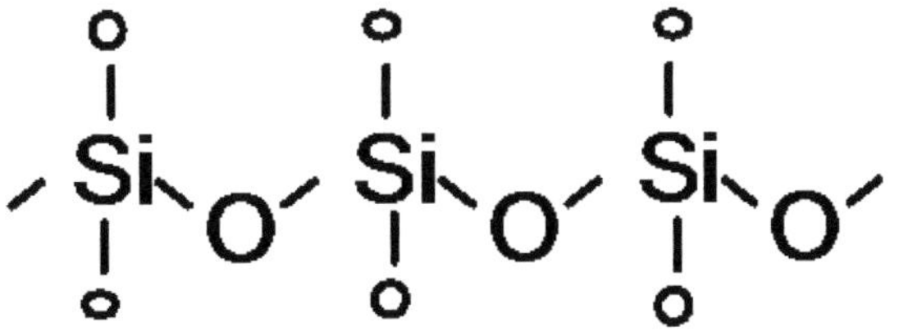

Fig. 9.14 Structure of silicones. The small open circles represent either CH_3 groups or other hydrocarbons. Source: Ref 9.1.

oxidation, and chemical attack. They have low toxicity. The uses of silicones include body implants, sealants for joints in construction, O-rings, and spatulas for cooking. Silicone greases for lubrication withstand higher temperatures than hydrocarbons.

Stretching of a Thermoplastic

When a linear polymer is stretched, the yielding occurs first in some region with the molecules becoming aligned with the direction of stretching. This results in a drop in stress (Fig. 9.15). Continued stretching occurs by propagation of the deformed region only after the whole specimen has yielded (Fig. 9.16).

NOTE OF INTEREST

Celluloid was the first synthetic plastic material. It was developed by John Wesley Hyatt in the late 1860s. He reacted a dispersion of cellulose nitrate and camphor. Celluloid is a tough material, with great tensile strength and resistance to water, oils, and dilute acids; it is capable of low-cost production in a variety of colors. Its uses included combs, collars, toys, film, and many other consumer goods. Although its flammability led to its replacement by newer synthetic polymers in many of these uses, celluloid has continued to be manufactured and widely used in Europe, the United States, Japan, and elsewhere.

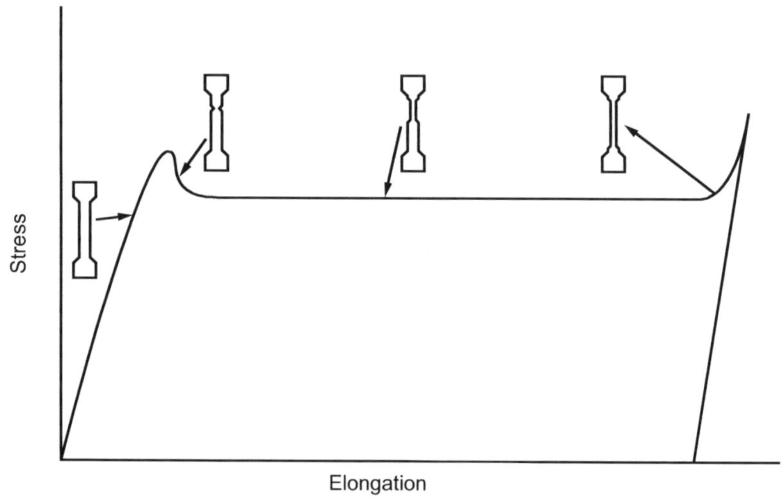

Fig. 9.15 Stress-strain curve for polyethylene. Note the formation and propagation of a necked region along the gage section. Source: Ref 9.2

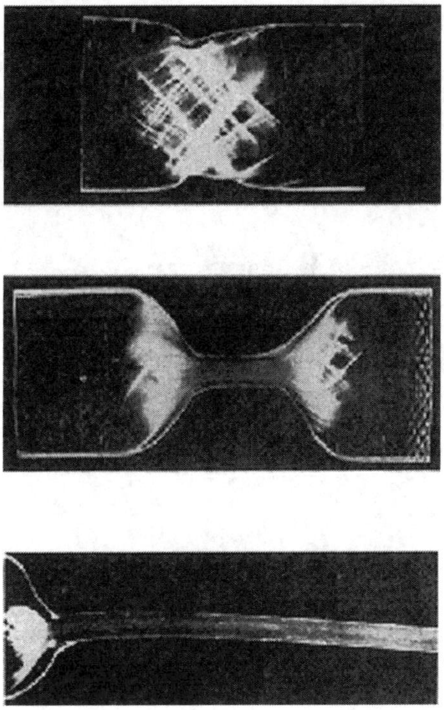

Fig. 9.16 Stages of neck formation and propagation in high-density polyethylene. Source: Ref 9.3

Exercises

1. **Expansion of Yielded Region of Plastic**

   Take the piece of plastic that holds a six-pack of beverage cans together and try to stretch the ligaments.

   This will require a high force at first, but then the force required will decrease as the yielded region expands. See Fig. 9.6.

2. **Decrease of Strength in 90° Stretching**

   After the plastic is stretched, try to stretch it at 90° to the original direction of stretching. Why does it come apart so easily?

   During the initial stretching, the molecules become aligned. This decreases the strength at 90° because only van der Waals bonds hold together the molecules.

3. **Shrinkage of Rubber Band**

   Stretch a rubber band by hanging a weight from it. Then heat the rubber band with a hot air blower. The weight will rise.

   The rubber band is shrinking as it is heated because the segments of chain between cross-links are vibrating. As they vibrate they shorten.

4. **Increasing Entropy in Rubber Band**

   Stretch a rubber band, then quickly release the tension and place it to your lips.

   It should feel cool because entropy (disorganization) increases and heat is now absorbed from your lips by the rubber band.

5. **Recycling Codes**

   The type of plastic often is identified by a recycling code. Identify as many of these as you can.

   Look on the bottom of polystyrene foam insulated cups, plastic soft drink bottles, and so forth.

REFERENCES

9.1 W.F. Hosford, *Materials for Engineers*, Cambridge University Press, 2008

9.2 W.F. Hosford, *Mechanical Behavior of Materials*, 2nd ed., Cambridge University Press, 2010

9.3 F.W. McClintock and A.S. Argon, *Mechanical Behavior of Materials*, Addison-Wesley, 1966

CHAPTER 10

Composites

WITH COMPOSITE MATERIALS, combinations of properties can be achieved that could not be achieved with individual materials because of the inclusion of reinforcing material. Some examples of composites are concrete, which is a composite of cement, sand, and gravel; steel-belted tires; plywood with alternating directions of fibers; carbon or glass fiber-reinforced polyester; and epoxy used for furniture, boats, and sporting goods. The reinforcing material of composites may be in the form of fibers, particles, or sheet laminates.

Fiber-Reinforced Composites

In fiber-reinforced composites, the reinforcing material is stronger and stiffer than the matrix material. Different geometric arrangements of the fibers are possible. The fibers may be unidirectionally aligned, aligned at 90° to one another in a woven fabric, or randomly oriented (Fig. 10.1). The fibers may be very long or chopped into short segments for easy fabrication. In thick sections, it is possible to have three-dimensional composites with randomly oriented short fibers. Fiber reinforcement is used to impart

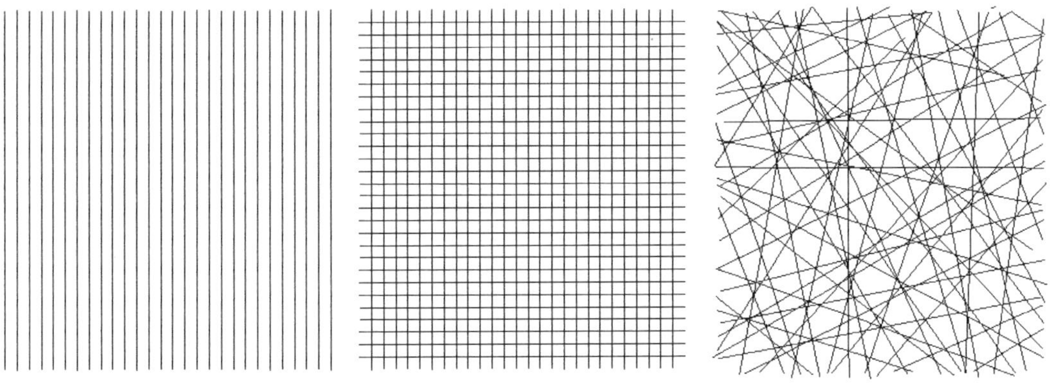

Fig. 10.1 Several geometric arrangements of fiber reinforcements. Source: Ref 10.1

stiffness (increased modulus of elasticity) or strength to the matrix. Fiber reinforcement also increases toughness.

The strains parallel to long parallel fibers must be the same in both the matrix and the fiber, $\varepsilon_f = \varepsilon_m = \varepsilon$. For loading parallel to unidirectionally aligned fibers, the elastic modulus is:

$$E = E_f V_f + E_m V_m$$

where E_f and E_m are the moduli of the fiber material and the matrix, and where V_f and V_m are the volume fractions of fiber and matrix. This often is called the *rule of mixtures*. It is an upper boundary to the elastic modulus of a composite. The modulus is very much lower for loading perpendicular to the fibers. Cross-plies or randomly oriented fibers give stiffening in other directions. A useful engineering approximation for randomly aligned fibers is:

$$E \approx (3/8)E_{\perp} + (5/8)E_{//}$$

where $E_{\perp}$ and $E_{//}$ are the moduli perpendicular and parallel, respectively, to unidirectional fibers.

The rule of mixtures cannot be used to predict the strengths of composites with uniaxially aligned fibers. The reason can be appreciated by considering the stress-strain behavior of both materials as shown schematically in Fig. 10.2. The strains in the matrix and fibers are equal, thus

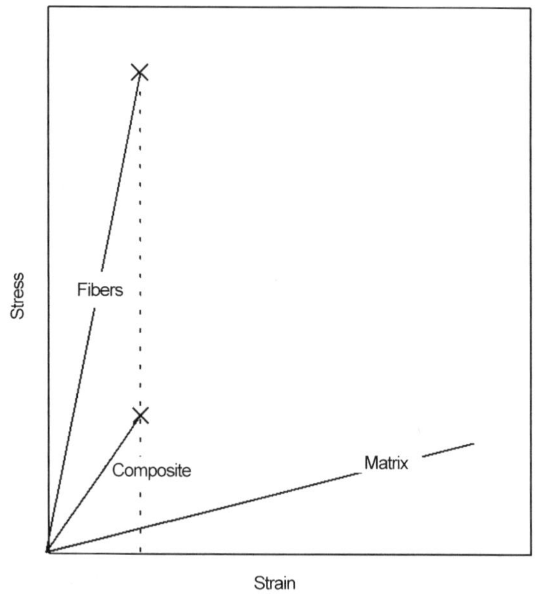

Fig. 10.2 Stress-strain curves for the matrix, the fibers, and the composite. Source: Ref 10.1

the fibers reach their breaking strengths long before the matrix reaches its tensile strength. Thus, the strength of the composite is $UTS < V_m UTS_m + V_f UTS_f$.

Usually the load carried by the fibers is greater than the breaking load of the matrix, so the composite will fail when the fibers break. The composite strength is given by:

$$UTS = V_m \sigma_m + V_f (UTS)_f$$

where $\sigma_m = (E_m/E_f)(UTS)_f$ is the stress carried by the matrix when the fiber fractures.

For composites with low-volume-fraction fibers, the fibers may break at a load less than the failure load of the matrix. In this case, after the fibers break, the whole load must be carried by the matrix, thus the predicted strength is $UTS = V_m(UTS)_m$.

Volume Fraction of Fibers

The stiffness and strength of reinforced composites should increase with the volume fraction of fibers, but there are practical limitations on the volume fraction. Fibers must be separated from one another. Fibers often are precoated to ensure this separation, and to control the bonding between fibers and matrix. Variability in fiber spacing (Fig. 10.3) may result during the infiltration of fiber arrangements by liquid resins. Approximately 55 to 60% is the practical upper limit for volume fraction fibers in unidirectional alignments. It is even lower in woven or cross-ply reinforcement.

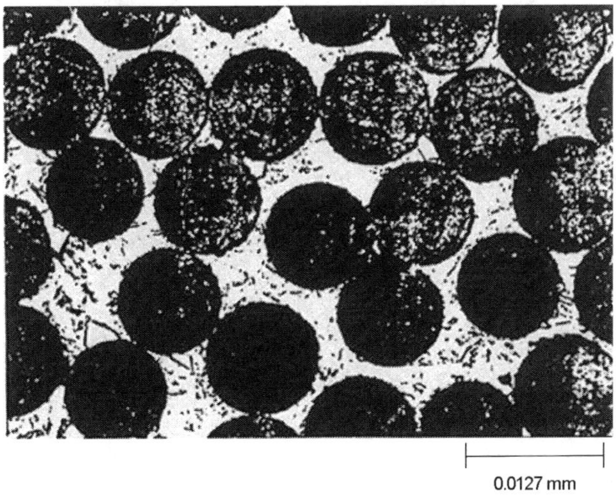

Fig. 10.3 Glass fibers in a polyester matrix. Note the variability in fiber spacing. Source: Ref 10.2

Fiber Length

Fabrication is much simplified if the reinforcement is in the form of chopped fibers. These can be blown onto a surface to form a mat. Shapes can be made with chopped fibers that are impossible with continuous fibers. Examples are extrusions, injection moldings, and transfer moldings. The disadvantage of chopped fibers is that some of the reinforcing effect of the fibers is sacrificed because the average axial stress carried by fibers is less for short ones than long ones. The average axial stress in a fiber depends on its aspect ratio, *D*/*L*, where *D* and *L* are the diameter and length of the fiber, respectively.

Failure may occur either by the fracture of fibers or by the fibers pulling out of the matrix. Both possibilities are shown in Fig. 10.4. Pullout will occur if the plane of the crack is near the end of the fiber. If it is not near the end, the fiber will fracture. Figure 10.5 is a picture showing the pullout of silicon carbide fibers in a titanium matrix.

More energy is absorbed if fibers pull out than if they break. The energy expended in fiber pullout increases with fiber length up to a critical length and then decreases with further length increase. Often greater toughness can be achieved with shorter fibers and lower fiber-matrix interface strength.

Epoxies and polyesters are common polymer matrices. Most of the polymers used for matrix materials have moduli of 2 to 3 GPa (290 to 435 ksi) and tensile strengths in the range of 35 to 70 MPa (5 to 10 ksi). Fiber rein-

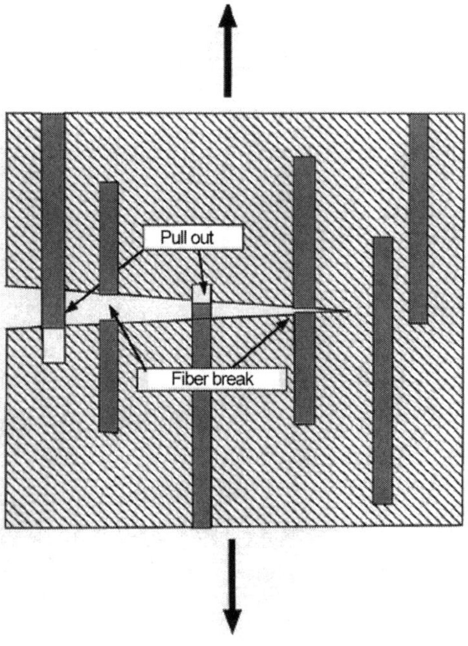

Fig. 10.4 Sketch showing some fibers fracturing at a crack and others pulling out. Source: Ref 10.1

forcements include glass, boron, aramid fiber, and carbon. Properties of some epoxy matrix composite systems are given in Table 10.1. Properties of some commonly used fibers are given in Table 10.2.

Other fiber composites include ceramics reinforced with metal or ceramic fibers. Metals such as aluminum-base alloys may be reinforced with ceramic fibers to increase their stiffness. In some eutectic systems,

Fig. 10.5 Photograph of SiC fibers pulling out of a titanium matrix. Source: Ref 10.3

Table 10.1 Properties of epoxy matrix composites

Fiber	Fiber, vol%	Young's modulus, GPa (ksi) Longitudinal	Young's modulus, GPa (ksi) Transverse	Tensile strength, MPa (ksi) Longitudinal	Tensile strength, MPa (ksi) Transverse
E-glass (unidirectional)	60	40 (5,801)	10 (1,450)	780 (113)	28 (4)
E-glass (bidirectional)	35	16.5 (2,393)	16.5 (2,393)	280 (40)	280 (40)
E-glass (chopped matte)	20	7 (1,015)	7 (1,015)	100 (14)	100 (14)
Boron (unidirectional)	60	215 (31,183)	24 (3,480)	1400 (203)	65 (9)
Kevlar29 (unidirectional)	60	50 (7,251)	5 (725)	1350 (195)	...
Kevlar49 (unidirectional)	60	76 (11,022)	6 (870)	1350 (195)	30 (4)
Carbon	62	145 (21,030)	...	1850 (268)	...

Kevlar is a registered tradename of E.I. du Pont de Nemours and Company.

Table 10.2 Typical fiber properties

Fiber	Young's modulus, GPa (ksi)	Tensile strength, MPa (ksi)	Elongation, %
Carbon (PAN* HS)	250 (36,259)	2.7 (0.39)	1.0
Carbon (PAN* HM)	390 (56,564)	2.2 (0.32)	0.5
SiC	70 (10,152)	...	...
Steel	210 (30,457)	2.5 (0.36)	...
E-glass	70 (10,152)	1.75 (0.25)	...
Boron	390 (56,564)	2.0–6.0 (0.29–0.87)	...
Kevlar29	65 (9,427)	2.8 (0.41)	4.0
Kevlar49	125 (18,129)	2.8 (0.41)	2.3
Al_2O_3	379 (54,969)	1.4 (0.20)	...
β-SiC	430 (62,366)	3.5 (0.51)	...

Kevlar is a registered tradename of E.I. du Pont de Nemours and Company.

NOTES OF INTEREST

The use of composites dates back to biblical times. Exodus chapter 5 describes how the Pharaoh forced the Israelites to use straw in making bricks.

The first use of reinforced concrete usually is attributed to Joseph-Louis Lambot in 1448. In 1868 Joseph Monier, a French gardener, patented a design for reinforced garden tubs and later patented reinforced concrete beams and posts for railway and road guardrails.

directional solidification can lead to rods of one phase reinforcing the matrix.

Particulate Composites

Composites reinforced by particles rather than long fibers are called particulate composites and include such diverse materials as concrete (cement matrix with sand and gravel particles) and "carbide tools" with a cobalt-base matrix alloy hardened by tungsten carbide particles.

Lamellar Composites

Two or more sheets of materials bonded together can be considered lamellar composites. Examples of lamellar composites include safety glass, plywood, plated metals, and glazed ceramics. The properties of plywood, such as stiffness and water expansion, are much less *anisotropic* (dependent on direction) than those of the wood itself.

Exercise

1. **Fractured Surface of Fiberglass**

   Break a piece of fiberglass and note the appearance of the fractured surface.

   Some glass fibers that have pulled out of the other surface should be visible, like those in Fig. 10.4.

REFERENCES

10.1 W.F. Hosford, *Mechanical Behavior of Materials*, Cambridge, 2005
10.2 *Composites*, Vol 1, *Engineered Materials Handbook*, ASM International, 1987
10.3 T.W. Clyne and P.J. Withers, *An Introduction to Metal Matrix Composites*, Cambridge University Press, 1993

CHAPTER 11

Wood

WOOD IS ONE OF THE FIRST MATERIALS used by man. Trees often are described as being either coniferous (cone bearing) or deciduous (losing leaves in the winter.) However, these descriptions are not mutually exclusive. Larch trees are both deciduous and coniferous, and holly trees are neither deciduous nor coniferous. Softwoods come from conifers, such as pine, spruce, and fir, and hardwoods come from deciduous trees, such as oak, maple, hickory, poplar, and willow. Most hardwoods are denser and harder than softwoods but there are exceptions. Douglas fir and eastern red cedar are harder and denser than willow and poplar.

Structure of Wood

Wood is composed of hollow cells, which trees use to transport sap. Annual growth rings are evident in the cross section of a tree trunk (Fig.

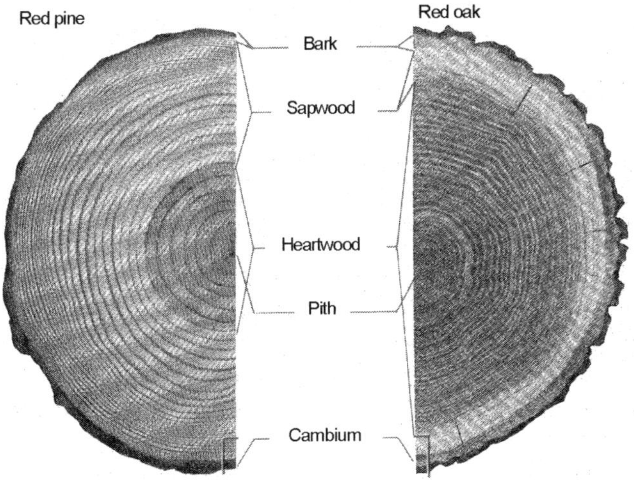

Fig. 11.1 Cross sections of typical softwood (red pine, left) and hardwood (red oak, right). Source: Ref 11.1

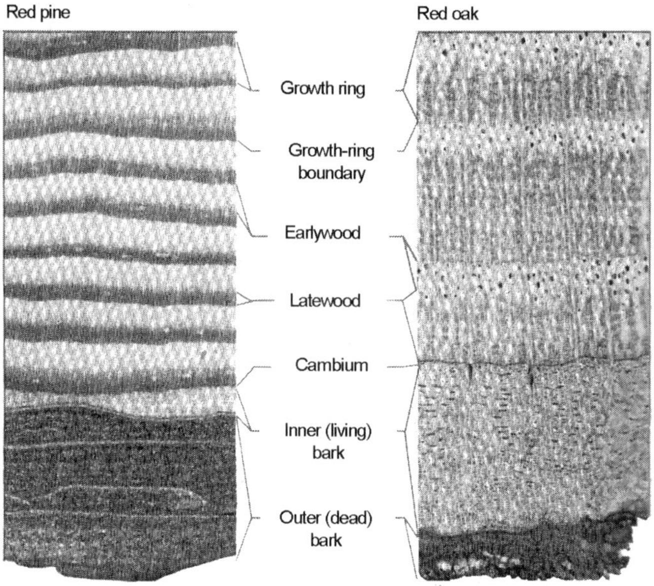

Fig. 11.2 Growth rings in typical softwood (red pine, left) and hardwood (red oak, right). Source: Ref 11.1

11.1). They are easily seen because faster growth in the spring results in larger cells (Fig. 11.2). The percentage of growth rings in the sapwood, the portion of the trunk though which sap is conducted, is between 20 and 40% for most hardwoods. Heartwood is dead sapwood and usually is darker in color.

Dimensional Changes with Moisture

Bone-dry wood consists mainly of three compounds: cellulose (40 to 50%), hemicellulose (15 to 25%) in the cell walls, and lignin (15 to 30%), which holds together the cells. In a living tree, wood usually contains approximately 30% moisture, thus wood shrinks as it dries. The shrinkage on drying varies with direction. Table 11.1 lists the shrinkage of several woods on drying to 6% moisture.

Because the shrinkage is greater in the tangential direction than in the radial direction, there is a strong tendency for radial cracks to form when a log is dried (Fig. 11.3). This tendency to split is aggravated because the outside dries before the inside does. To prevent this, lumber usually is cut from logs before oven drying to approximately 6% moisture.

Even after being sawn into board, the directional dependence of shrinkage causes warping, as illustrated in Fig. 11.4.

After drying, the moisture content of wood will approach equilibrium with the atmosphere. At 20 °C (70 °F), the moisture content may vary

Table 11.1 Percent shrinkage of woods on drying

	Shrinkage, %	
Wood	Radial direction	Tangential direction
White pine	2.1	6.1
Douglas fir	4.8	7.6
Western cedar	2.2	4.9
Yellow birch	7.3	9.5
Hickory	4.9	8.9
Maple	3.7	7.1
White oak	5.6	10.5

Note: In all cases the longitudinal shrinkage is less than 0.25%.

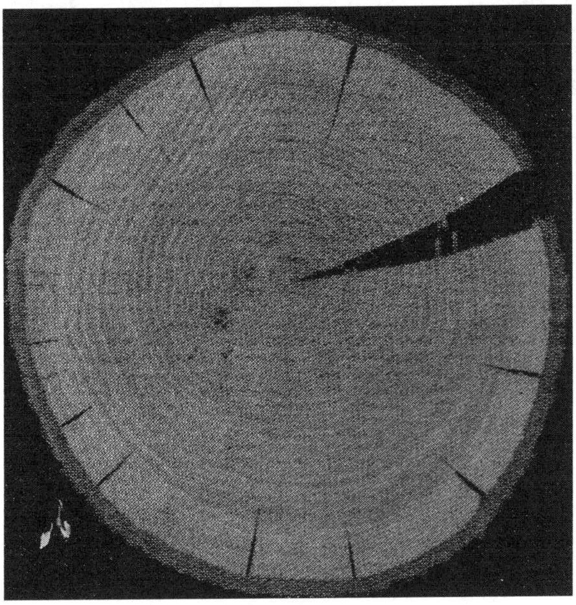

Fig. 11.3 Splitting of a red oak log on drying. Source: Ref 11.1

from 4.5% if the humidity is 20%, to 16% if the humidity is 80%. The dimensional changes that result from a change in moisture content vary with direction, as listed in Table 11.2.

It is interesting to note that one often can tell whether a picture frame was made in the winter or the summer (Fig. 11.5). In the winter, the 45° angles at the corners will have decreased in angle if the frame was made in more humid weather (summer), thus the corners will open on the *inside* (Fig. 11.5A). On the other hand, if the frame was made in the winter (low humidity), the 45° angles at the corners will open on the *outside* in the summer (Figure 11.5B).

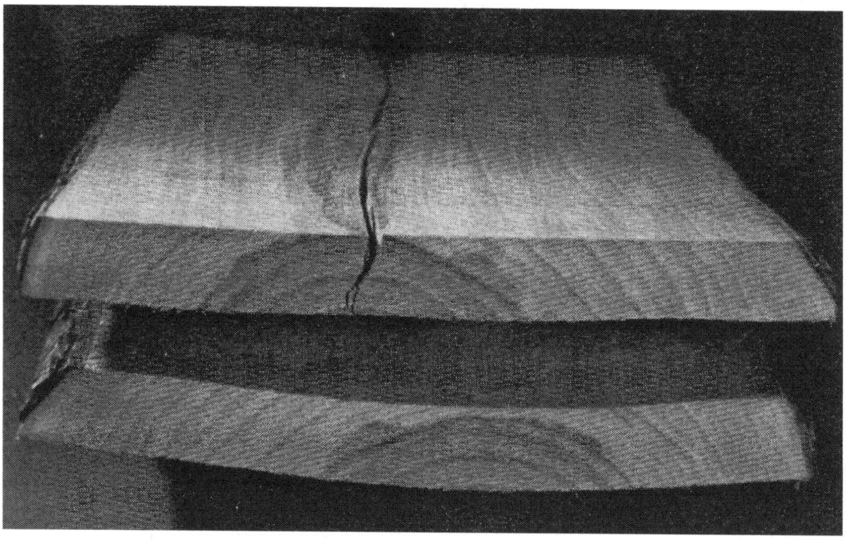

Fig. 11.4 Elm board warped because the shrinkage in the tangential direction is greater than in the radial direction (bottom). Splitting when a warped board is flattened (top). Source: Ref 11.1

Table 11.2 Dimensional change in wood with change of moisture content

Direction	Dimensional change per 1% moisture change
Radial	0.15%
Tangential	0.25%
Longitudinal	0.01%

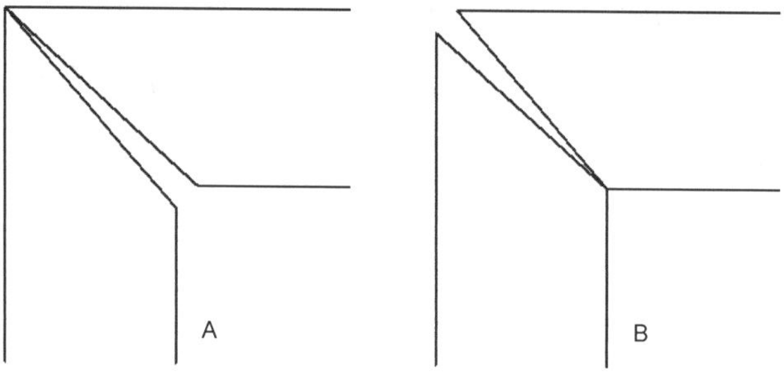

Fig. 11.5 A picture frame made in the summer will open on the inside during the winter (A), while a picture frame made in the winter will open on the outside during the summer (B). Source: Ref 11.2

Anisotropy of Properties

Table 11.3 shows that the elastic modulus and tensile and compressive strengths vary with direction.

Plywood

The orientation dependence of properties can be largely circumvented by the use of plywood, composed of plies oriented with their grain at 90° to one another. The plies are cut from rotating logs that have been softened with moisture (Fig. 11.6). The number of plies is always an odd number to prevent warping as changes occur in moisture content.

Paper and Cardboard

The earliest paper was papyrus made from reeds by the ancient Egyptians. Today, paper is made from pulp consisting of ground up wood in water. Often chemicals are added to help decompose the wood. The pulp is heated to soften the fibers. Some clay or cloth may be added. Clay is used to produce shiny paper used in magazines. Newsprint and cardboard

Table 11.3 Properties of common woods at 12% moisture content

Property	White oak	Douglas fir
Density, Mg/m^3	0.68	5.0
Young's modulus, GPa (ksi)	6.0 (0.87)	2.9 (0.42)
Tensile strength	(a)	(a)
Radial, MPa (ksi)	5.4 (0.8)	2.4 (0.3)
Compressive strengths, MPa (ksi)		
Longitudinal	51 (36)	51 (36)
Radial	7.4 (1.1)	5.2 (0.8)

(a) The longitudinal tensile strengths are approximately 20 times the radial tensile strengths.

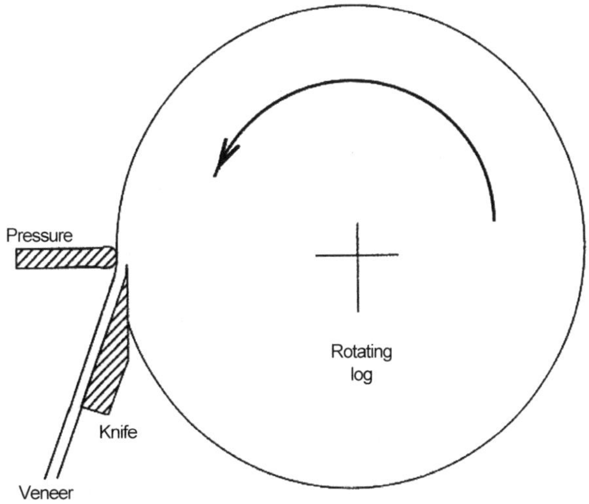

Fig. 11.6 Cutting a log for plywood and veneer. Source: Ref 11.2

NOTE OF INTEREST

The plywood industry began in 1907, when the Portland Manufacturing Company installed an automatic glue spreader and a hand press to produce plywood. They made 420 panels a day. Initially the market was primarily for door panels. Automobile running boards were made from plywood but this stopped because of water damage. In 1934, Dr. James Nevin, a chemist at Harbor Plywood Corporation in Aberdeen, Washington, finally developed a fully waterproof adhesive for plywood. This technological advancement had the potential to open up significant new markets. After World War II, the industry produced 1.4 billion square feet of plywood. By 1954, the industry approached 4 billion square feet. By 1975, U.S. production alone exceeded 16 billion square feet.

contain no clay. Water is removed by squeezing—often by rolling, which tends to align the fibers, making the properties very anisotropic.

Exercises

1. **Dimensional Increases of Wood upon Hydration**

   Take a small piece of wood and measure its dimensions. Immerse it in water for a day, then remove the wood and measure the dimensions again. Which dimension increased the most? Which the least? Why?

   The greatest dimensional increase will be in the direction that is perpendicular to the grain (the direction that was parallel to the circumference of the tree.) The least dimensional increase will be parallel to the grain.

2. **Paper Fiber Alignment**

   Take a piece of newspaper and try to make a straight tear down the page. Then try to make a straight tear across the page. Explain the difference and deduce the direction in which the paper was rolled.

   Rolling during the manufacture of the paper aligns the fibers. Paper will tear most easily in this direction.

3. **Cardboard Fiber Alignment**

   Cut a square piece of cardboard (20 × 20 cm, or 8 × 8 in.) from the back of a pad of paper. Suspend it as shown in Fig. 11.7 and place a small weight on it. Note how much the cardboard deflects. Then turn the cardboard 90° and repeat. Explain the difference and deduce the direction in which the cardboard was rolled.

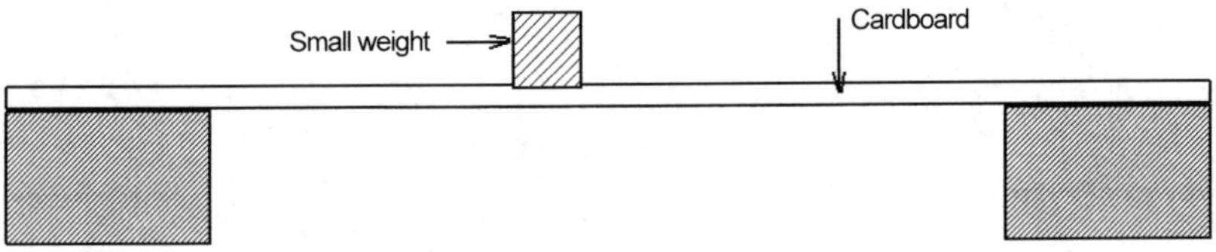

Fig. 11.7 Setup for Exercise 3

The cardboard will deflect least when the direction it was rolled is the direction between the supports.

REFERENCES

11.1 R.B. Hoadley, *Understanding Wood*, The Taunton Press, 1980
11.2 W.F. Hosford, *Materials for Engineers*, Cambridge University Press, 2008

CHAPTER 12

Corrosion

THE COST OF CORROSION in the United States is estimated to be \$1 trillion annually. Approximately 40% of this cost can be avoided by well-known practices. Much of this loss is due to the corrosion of iron and steel, although many other metals also corrode. There are two principal types of corrosion: aqueous corrosion (corrosion in water) and direct corrosion (attack in air).

Corrosion in Water Solutions

Aqueous corrosion always involves an electrolytic cell with a cathode and an anode (Fig. 12.1). It is the opposite of electroplating. The anode is where ions go into solution $M \rightarrow M^{+n} + ne^-$. The cathode is where electrons are consumed. Possible cathode reactions are:

$M^{+n} + ne^- \rightarrow M$ (This can occur only if there is a high concentration of M^{+n} ions.)

$2H^+ + 2e^- \rightarrow H_2$ (This can occur only if the solution is acid.)

$O_2 + 2H_2O + 4e^- \rightarrow 4(OH)^-$ (There must be O_2 in solution. This is the most common cathode reaction.)

$O_2 + 4H^+ + 4e^- \rightarrow 2H_2O$ (There must be O_2 in an acid solution.)

Table 12.1 shows the electromotive series. It gives the electrode potentials for various possible half reactions for solutions with 1 molar concentration.

The voltages for the various reactions are measured against a hydrogen electrode $2H^+ + 2e^- \rightarrow H_2$, which is defined as zero (Fig. 12.2).

The electrode potentials are different in other aqueous solutions. Seawater is a common corrosive environment. The galvanic series ranks alloys according to their cathode potential in seawater (Table 12.2).

The potential depends on the concentration. Often cells arise as a result of different concentrations in different places. Oxygen concentration cells are very common. Figure 12.3 illustrates some sources of oxygen concen-

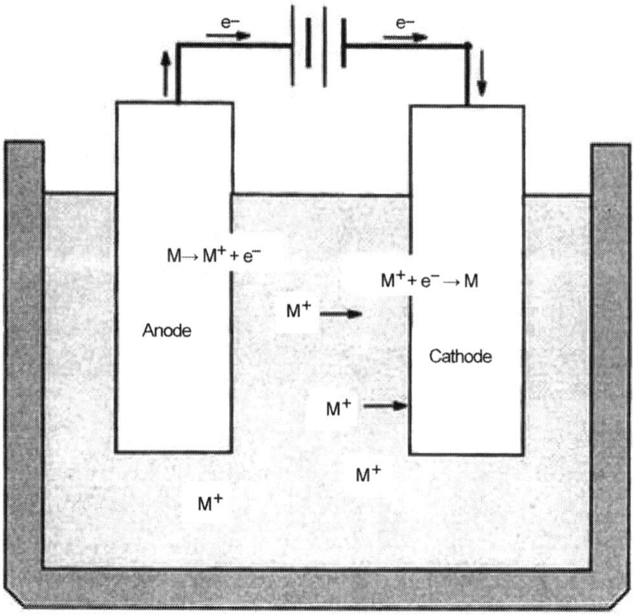

Fig. 12.1 Corrosion cell. The anode is where electrons are generated in the external circuit, and the cathode is where they are consumed. Source: Ref 12.1

Table 12.1 Electromotive series

Anode half-cell reaction (the arrows are reversed for the cathode half-cell reaction)	Electrode potential used by electrochemists and corrosion engineers, volts	
$Au \rightarrow Au^{3+} + 3e^-$	+1.50	
$2H_2O \rightarrow O_2 + 4H^+ + 4e^-$	+1.23	↑ Cathodic (noble)
$Pt \rightarrow Pt^{4+} + 4e^-$	+1.20	
$Ag \rightarrow Ag^+ + e^-$	+0.80	
$Fe^{2+} \rightarrow Fe^{3+} + e^-$	+0.77	
$4(OH)^- \rightarrow O_2 + 2H_2O + 2e^-$	+0.40	
$Cu \rightarrow Cu^{2+} + 2e^-$	+0.34	
$H_2 \rightarrow 2H^+ + 2e^-$	0.000	Reference
$Pb \rightarrow Pb^{2+} + 2e^-$	−0.13	
$Sn \rightarrow Sn^{2+} + 2e^-$	−0.14	
$Ni \rightarrow Ni^{2+} + 2e^-$	−0.25	
$Fe \rightarrow Fe^{2+} + 2e^-$	−0.44	Anodic (active) ↓
$Cr \rightarrow Cr^{2+} + 2e^-$	−0.74	
$Zn \rightarrow Zn^{2+} + 2e^-$	−0.76	
$Al \rightarrow Al^{3+} + 3e^-$	−1.66	
$Mg \rightarrow Mg^{2+} + 2e^-$	−2.36	
$Na \rightarrow Na^+ + e^-$	−2.71	
$K \rightarrow K^+ + e^-$	−2.92	
$Li \rightarrow Li^+ + e^-$	−2.96	

Source: Ref 12.2

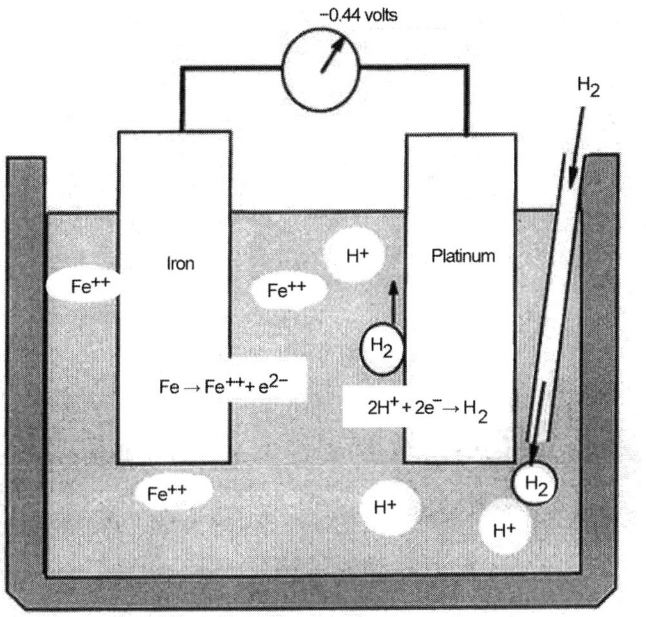

Fig. 12.2 The hydrogen cathode

Table 12.2 The galvanic series of common alloys

Cathodic ↑	↓ Anodic
Graphite	Nickel—A
Silver	Tin
12% Ni, 18% Cr, 3% Mo steel—P	Lead
20% Ni, 25% Cr steel—P	Lead–tin solder
23 to 30% Cr steel—P	12% Ni, 18% Cr, 3% Mo steel—A
14% Ni, 23% Cr steel—P	20% Ni, 25% Cr steel—A
8% Ni, 18% Cr steel—P	14% Ni, 23% Cr steel—A
7% Ni, 17% Cr, steel—P	8% Ni, 18% Cr steel—A
16 to 18% Cr steel—P	7% Ni, 17% Cr steel—A
12 to 14% Cr steel—P	Ni-resist
80% Ni, 20% Cr—P	23 to 30% Cr steel—A
Inconel—P	16 to 18% Cr steel—A
60% Ni, 15% Cr—P	12 to 14% Cr steel—A
Nickel—P	4 to 6% Cr steel—A
Monel metal	Cast iron
Copper–nickel	Copper steel
Nickel–silver	Carbon steel
Bronzes	Aluminum alloy 2017-T
Copper	Cadmium
Brasses	Aluminum, 1100
80% Ni, 20% Cr—A	Zinc
Inconel—A	Magnesium alloys
60% Ni, 15% Cr—A	Magnesium

Note: A, active state; P, passive state. Source: Ref 12.2

tration cells. The role of oxygen in corrosion at times can seem anomalous. Oxygen normally is required for corrosion. However, the metal loss does not occur where the oxygen concentration is the greatest.

Corrosion cells can arise from two-phase microstructures, from cold work (Fig. 12.4), and from grain boundaries (Fig. 12.5).

Accumulation of positive ions near the anode and negative ions near the cathode (Fig. 12.6) will polarize the cell, decreasing the voltage. With polarization, the corrosion current decreases, as indicated in Fig. 12.7.

Passivation

With some materials, when the anode potential reaches a critical value, the corrosion current drops abruptly to a very low rate (Fig. 12.8). This condition is called *passivation*. A very thin oxygen layer on stainless steels is

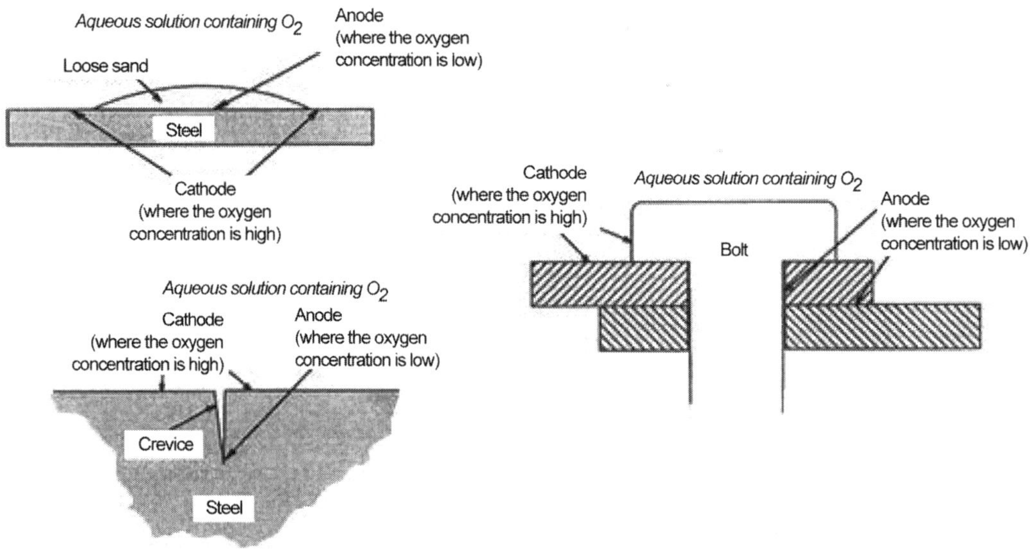

Fig. 12.3 Oxygen concentration cells. The regions that are shielded from oxygen are the anodes, while the cathode reaction occurs where oxygen is plentiful.

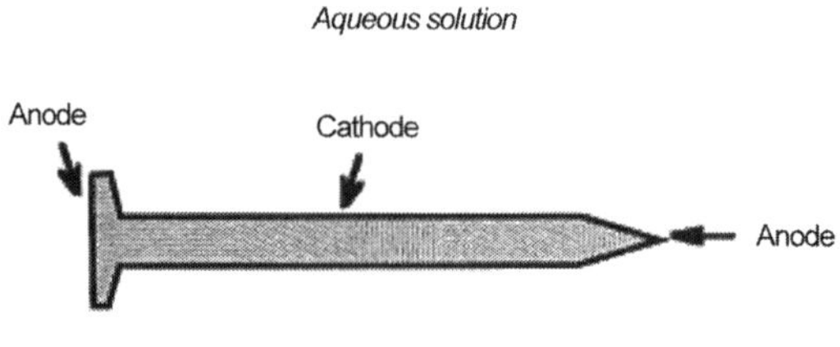

Fig. 12.4 Regions that have been cold worked are anodic to regions that have not.

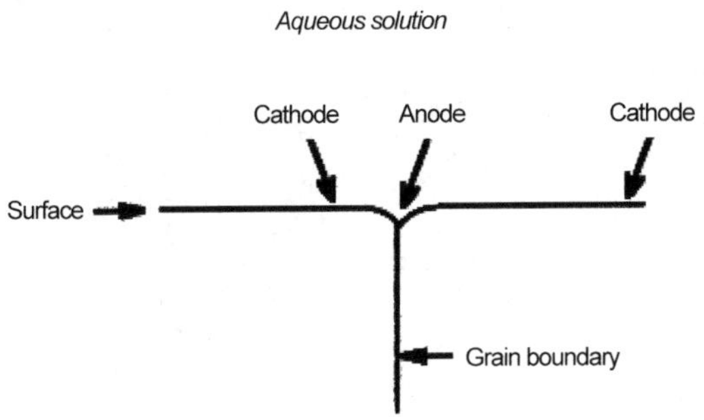

Fig. 12.5 Because atoms at grain boundaries are in a higher energy state, the grain boundaries become anodic.

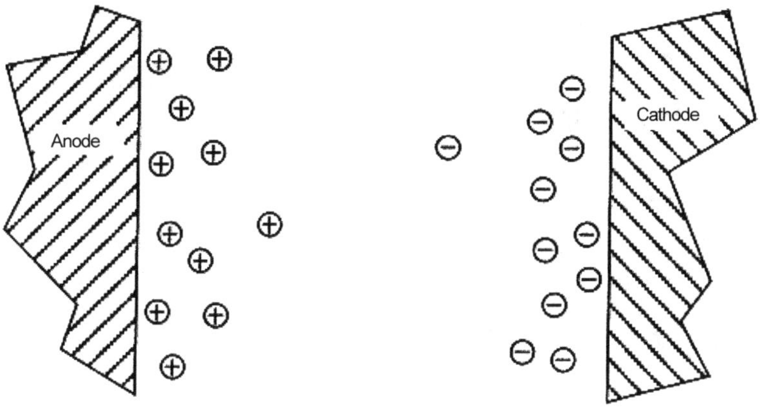

Fig. 12.6 Polarization is caused by an accumulation of positive ions near the anode and negative ions near the cathode.

sufficient to cause passivation. Oxygen and a very small amount of corrosion are required to maintain the passive state. In Table 12.2 (the galvanic series of common alloys) stainless steels occupy two places, depending on whether they are passive or not. Titanium alloys may be passive under special conditions.

If conditions are oxidizing enough, even carbon steel can become passive. Concentrated nitric acid can be shipped in mild steel containers. However, the containers would corrode extremely rapidly if filled with dilute nitric acid.

Oxygen is required for passivity of stainless steels. In one case, simply identifying stainless steel pipes by attaching transparent adhesive tape to the outside of the pipes was responsible for corrosion when moisture condensed on them. The tape prevented the access of oxygen so the regions under the pipe lost their passivity.

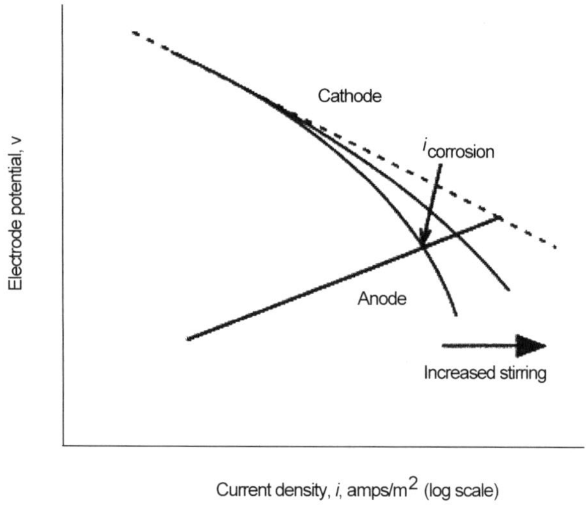

Fig. 12.7 Polarization at the cathode decreases the cell potential. Increased convection decreases the polarization. The effects of anode polarization are similar.

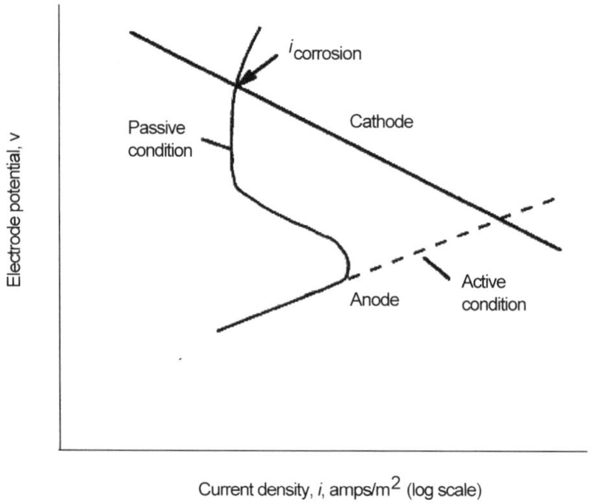

Fig. 12.8 Above a critical anode potential, certain materials become passive. Their corrosion rate drops abruptly. Note the current density is plotted on a logarithmic scale.

Corrosion Control

The means of controlling corrosion are means of disrupting or changing the corrosion cell. One way is to remove the electrolyte, because corrosion will not occur in the absence of water. Inert coatings such as paint serve this function. Another is to break the circuit. Cells caused by electrical contact of dissimilar metals can be interrupted by placing an insulator between the metals. Reversing the voltage by imposing opposite voltage with external circuit or sacrificial corrosion (Fig. 12.9) are other means. With galvanized steel, the zinc is anodic to iron and hence protects the steel by sacrificial corrosion versus inert coatings (Fig. 12.10).

Stress Corrosion

Stress corrosion can occur in some metals when loaded in tension in a specific environment. Cracks run through grains and along grain boundaries. Brass is susceptible to stress corrosion in the presence of ammonia. Chloride ions form an environment that leads to stress corrosion of many metals.

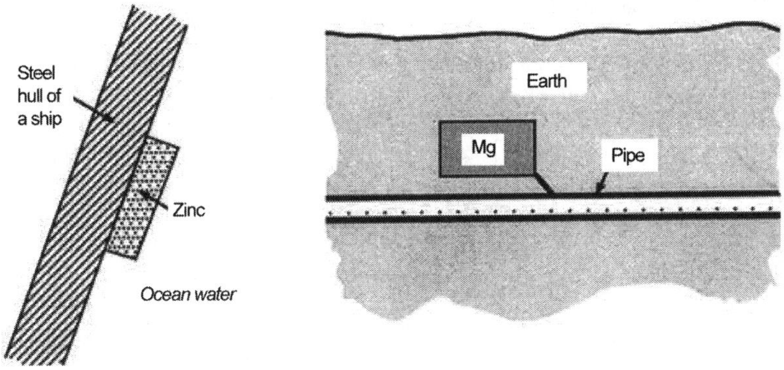

Fig. 12.9 Corrosion protection by sacrificial corrosion of zinc (left) and magnesium (right)

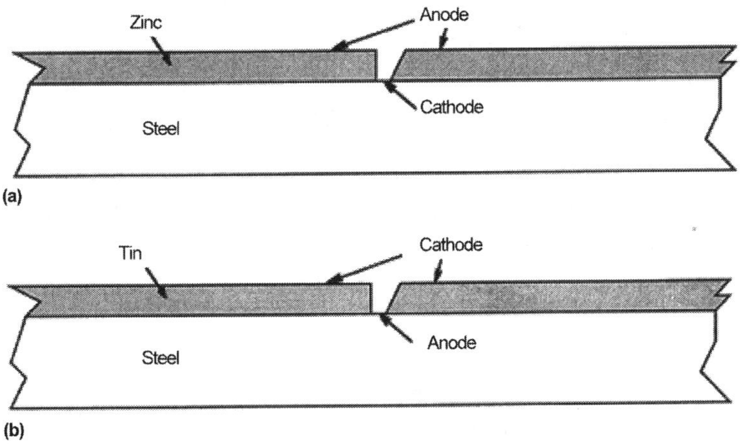

Fig. 12.10 Aqueous solution. (a) Plating steel with zinc (galvanizing) offers cathodic protection to steel if the plating is scratched. (b) Tin plating offers no cathodic protection so the steel will corrode if the plating is scratched.

Rust

Rust is ferric oxide, Fe_2O_3, and ferric hydroxide, $Fe(OH)_3$. Ferrous ion, Fe^{2+}, is soluble, but further oxidation produces ferric ions, $3Fe^{2+} + 6OH^- \rightarrow 3Fe(OH)_3 + 3H_2O$. Ferric hydroxide is insoluble and precipitates as rust. If dried, the ferric hydroxide turns into an oxide, $2Fe(OH)_3 \rightarrow Fe_2O_3 + 3H_2O$. Often, the rust-producing reaction occurs at some distance from where the anode reaction occurs, so rust deposits may not be directly over the corroded region. This is illustrated in Fig. 12.11.

Special steels containing small amounts of copper and nickel form more protective rusts. These steels do rust, but more slowly than ordinary steels. This has led to their use in architecture and sculpture.

Direct Oxidation

Direct oxidation of most metals in the absence of water occurs only at high temperature. It might seem that high-temperature oxidation would not involve an electrolytic cell. However, there is an anode and cathode. The anode reaction is $M \rightarrow M^{n+} + ne^-$, and the cathode reaction is $O_2 + 4e^- \rightarrow 2O^{2-}$. Either O^{2-} ions or M^{n+} ions and e^- must diffuse through the oxide. The M^{n+} ions are smaller than O^{2-} ions and therefore diffuse faster. Hence, their diffusion is rate controlling. Figure 12.12 illustrates the reactions and transport in direct oxidation.

Diffusion and electron transport through Al_2O_3 and Cr_2O_3 are very slow because they are very free of defects. Hence, they are very protective. For an oxide to be protective, it must cover the surface. Hence, volume of oxide $\geq$ volume of metal oxidized. If $nM + mO \rightarrow pM_nO_m$, protection occurs if:

$$(MW)_{oxide}/\rho_{oxide} \geq n(AW)_M/\rho_M$$

The equation is not satisfied for sodium, potassium, alkali, and the alkaline earth metals, so they oxidize rapidly in air. If the ratio $\left[(MW)_{oxide}/\rho_{oxide}\right]/\left[n(AW)_M/\rho_M\right]$ is too high, the compressive stresses in the oxide may cause it to spall off. This is the case for Fe_2O_3. Another necessary condition for a protective oxide is that it must be solid. Tung-

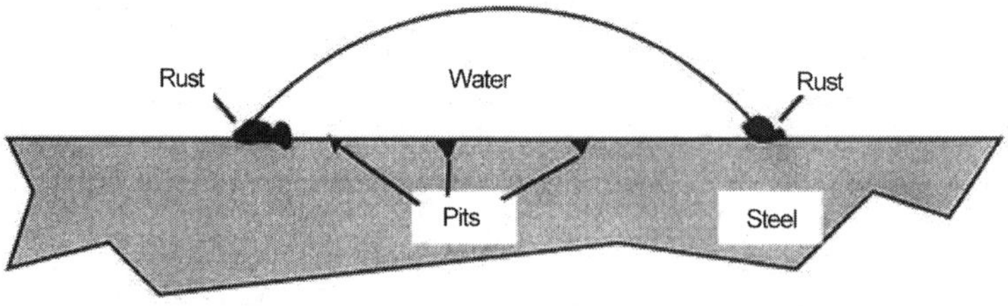

Fig. 12.11 Rust formation away from corroded sites where the oxygen concentration is higher

sten and molybdenum oxidize very rapidly at high temperatures because their oxides are volatile. Because vanadium pentoxide (V_2O_5) forms a low-melting eutectic with Fe_2O_3, which flows off of the surface, fuels containing a small amount of vanadium have caused serious problems in power-generating turbines.

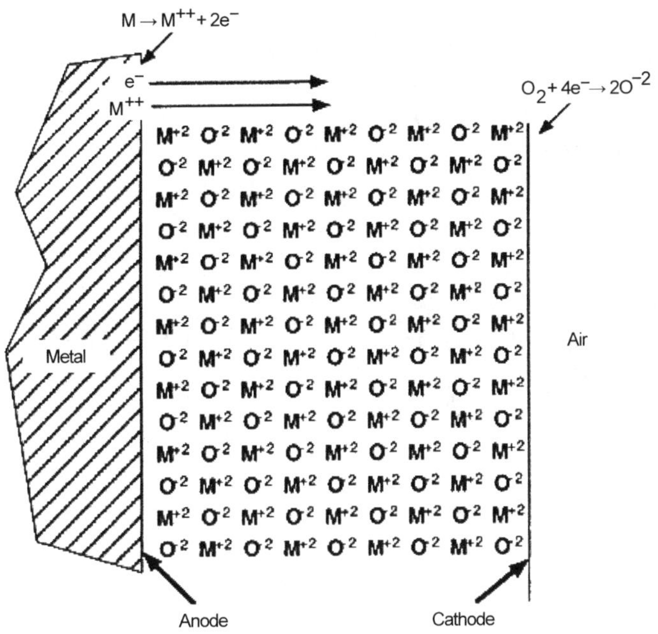

Fig. 12.12 Direct oxidation. Oxide forms by diffusion of anions and electrons to the oxide-air surface.

NOTES OF INTEREST

Francis Wollaston in about 1815 first suggested that acid corrosion was electrochemical. Sir Humphrey Davy in 1824 showed that there was a voltage between two different metals when connected and immersed in water. From this work he showed that a voltage was set up between them and suggested that the copper bottoms of ships could be protected by attaching iron or zinc plates to them. In about 1830, Theodore Sturgeon showed that there could be potentials between different regions on the same metal surface. He found that differences of electrolyte concentration and temperature caused potential differences.

At the time of WWI, the U.S. Army did not specify that brass cartridge cases be given stress-relief anneals. This caused a problem when cow barns in France were used to store ammunition. As anyone who has changed diapers knows, ammonia is one of the decomposition products of urine. Ammonia from cow urine caused stress-corrosion cracking of the brass cartridge cases that contained residual tensile stresses in the hoop direction. Since then, the Army has specified stress relief annealing.

Exercise

1. **Corrosion of Steel Wool**

   Put pads of steel wool into two jars of water. Boil one jar and securely cap it. Leave the other open to the air. Compare the two jars after two days. Explain the difference.

   Corrosion of the steel wool will be observed in the open jar but not in the capped jar. The boiling removes dissolved oxygen. There can be no corrosion without oxygen, thus no corrosion can occur in the capped jar. Oxygen from the air will dissolve in the open jar, allowing corrosion.

REFERENCES

12.1 W.F. Hosford. *Physical Metallurgy*, 2nd ed., CRC Press, 2010
12.2 L.H. Van Vlack, *Elements of Materials Science and Engineering*, 3rd ed., Addison-Wesley, 1974

CHAPTER 13

Forming and Shaping

Liquid to Solid Processing

Metals usually are cast into molds of sand, another refractory material, or another metal. Steel and iron are cast into sand or refractory molds. Gold, silver, and refractory metals are cast into molds of high-temperature ceramics. Aluminum may be cast into sand or iron molds. Low-melting-point metals (aluminum, magnesium, and zinc) may be *die cast*. In die casting, metal is forced into a metal mold that automatically opens and ejects the part as soon as it is frozen. The production rate is high but so are capital costs. In recent variations of this process, called *thixotropic* or *rheocasting*, the melt is approximately half solid-half liquid when it is injected into the mold.

Injection molding of thermoplastics is very much like injection molding of metals. Thermoplastics also can be extruded or blow molded.

Single crystals may be grown by directional solidification. With the Bridgman technique, a mold is slowly removed from a furnace (Fig. 13.1). Freezing starts at one end and slowly progresses to the other. Because

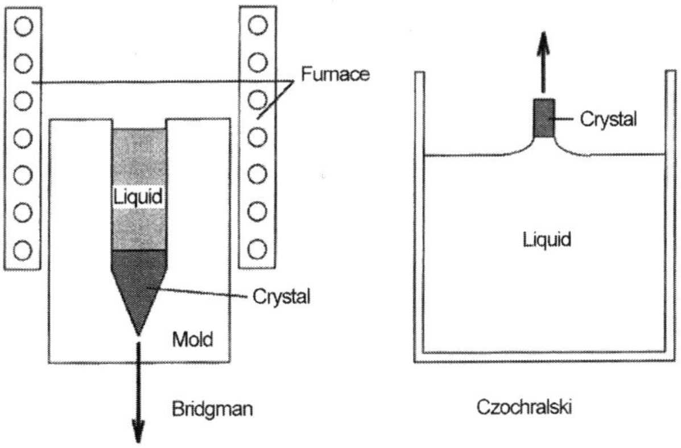

Fig. 13.1 Bridgman and Czochralski methods of growing single crystals. Source: Ref 13.1

freezing starts at a point, only one crystal is nucleated. The Czochralski method involves lowering a seed crystal into the melt so that it partially melts, and then withdrawing it slowly upward. The growing crystal is rotated about a vertical axis to help stir the liquid. The lack of a mold eliminates contamination from mold walls but makes it impossible to control the exterior shapes of the crystals. This is the method used to grow silicon crystals for the semiconductor industry.

Bulk Forming of Solid Metals

Bulk forming of cast ingots can be accomplished by rolling, extrusion, forging, and drawing. The important forces are compressive. Figure 13.2 illustrates several bulk forming processes.

Hot Working versus Cold Working

Many bulk forming processes are done hot. At high temperatures lower tool forces are required. Consequently, equipment size and power requirements are decreased. *Hot working* often is defined as working above the recrystallization temperature so that the work metal recrystallizes as it deforms in the time period between repeated operations, as in forging and

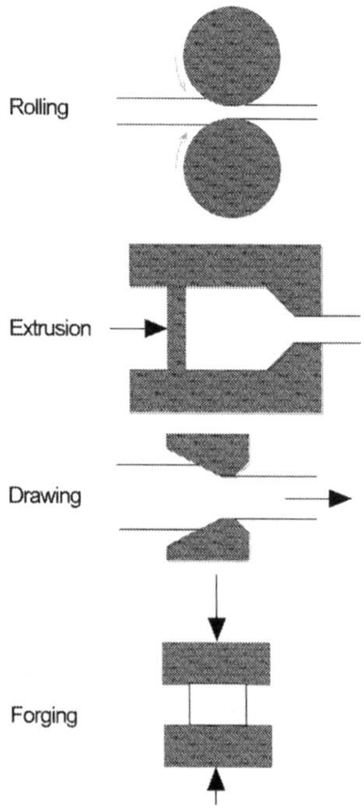

Fig. 13.2 Several bulk forming processes

multiple-stand rolling, or while the material is cooling to room temperature after the deformation is complete. High temperatures lower the flow stress, whether recrystallization occurs during the deformation or not. Hot rolled products are in an annealed state.

The elevated temperatures during hot working have several undesirable effects. Among them are:

- Lubrication is more difficult. Although viscous glasses often are used in hot extrusions, most hot working is done without any lubrication.
- The work metal tends to oxidize. Scaling of steel and copper alloys causes loss of metal and roughened surfaces. Although processing under inert atmosphere is possible, it is prohibitively expensive and is avoided except in the case of very reactive metals, such as titanium.
- Tool life is shortened because of heating, the presence of abrasive scales, and the lack of lubrication.
- Poor surface finish and loss of precise gage control result from the lack of adequate lubrication, oxide scales, and roughened tools.
- The lack of work hardening is undesirable where the strength level of a cold worked product is needed.

Because of these limitations, it is common to hot roll steel to a thickness of approximately 2 mm (0.08 in.) to take advantage of the decreased flow stress at high temperature. The hot rolled product then is pickled to remove scale. Further rolling is done cold to ensure good surface finish and optimal mechanical properties. The cold rolled steel sheet almost always is sold in an annealed state. Annealing is done in a controlled atmosphere after cold rolling. The principal advantage of *cold rolled* sheet steel over hot rolled sheet is a much better surface finish.

Sheet Forming

In sheet forming processes the important stresses are tensile. Sheets can be formed into complex shapes by stamping. In stampings the sheets are clamped at their edges and stretched to a final shape by a shaped punch (Fig. 13.3). Draw beads control how much material from the flange is drawn into the die. Wrinkling of the finished part results if too much material is allowed to flow into the die. It should be emphasized that in stamping processes the sheet is not squeezed between the die and punch. Rather, tensile forces in the sheet cause it to deform by stretching. If too much stretching occurs, the sheet may suffer a tensile failure in the die.

Cups can be formed by deep drawing discs cut from sheets (Fig. 13.4). Cups and cans are made by deep drawing. The maximum diameter reduction is limited by failure of the cup walls. This usually limits the height-to-diameter ratio to approximately 0.75. Deeper cups can be made by redrawing and *ironing* (thinning of the walls).

Polymer Processing

Thin sheets and films to form bags are made from thermoplastics by blowing air into an extruded tube. The air expands the tube, which then is collapsed by nip rollers. The thickness and size of the film is controlled by the volume of air inside the bubble, the speed of the nip rollers, and the output rate of the extruder.

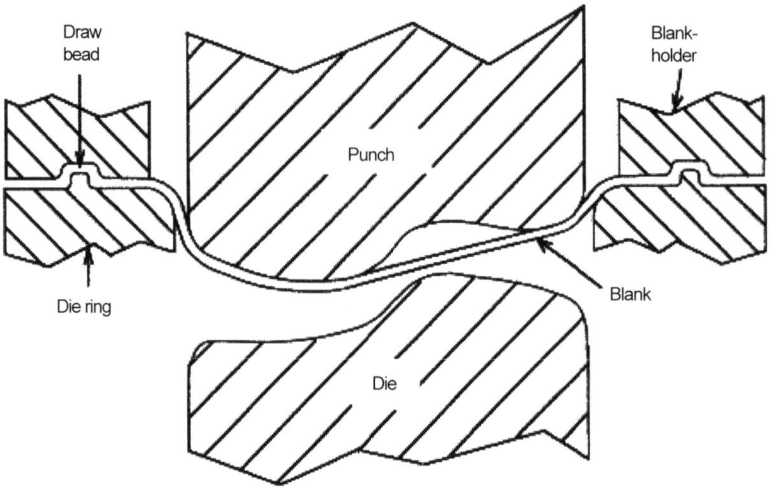

Fig. 13.3 Stamping of a sheet metal part. Note that the restraint from the draw beads in the blank holder promotes more stretching of the sheet. Source: Ref 13.2

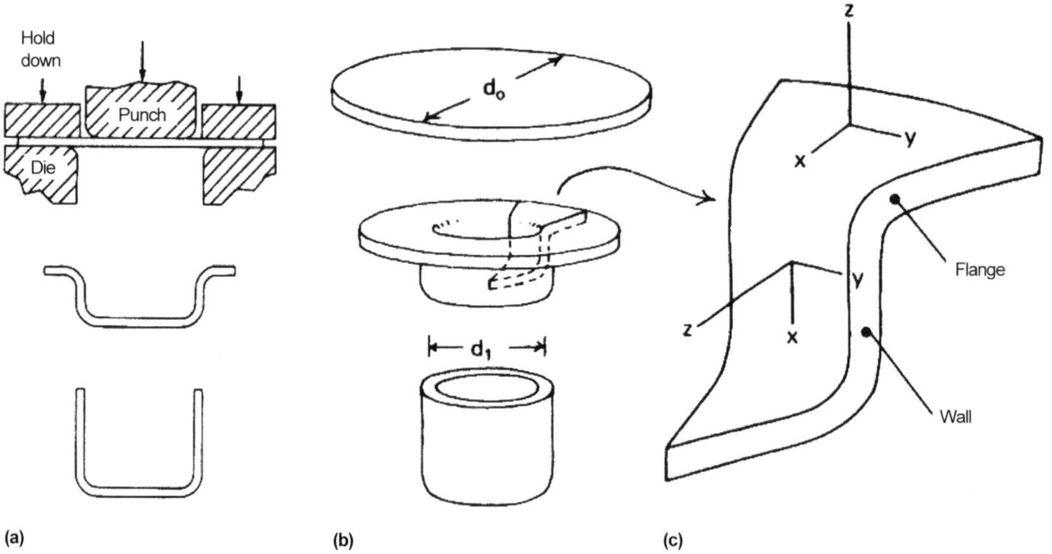

Fig. 13.4 Schematic illustration of forming a cup from sheet metal. (a) A descending punch forces the sheet through a circular hole in the die. A hold-down plate prevents buckling. (b) The major deformation is the circumferential contraction of the flange so that it can pass over the die lip. (c) By focusing on a pie-shaped segment, it can be seen that the flange is under hoop compression and the wall under tension. Source: Ref 13.1

Plastic bottles are made by blow molding. The process begins with the conventional extrusion of a *parison* or closed-end tube. The parison then is clamped inside a hollow mold and inflated. The air pressure forces the parison against the mold surface until it cools in the shape of the interior of the mold cavity. The mold then is opened and the plastic bottle is ejected. Blow molding is used with many different plastics, including high-density polyethylene (HDPE), polyvinyl chloride (PVC), polycarbonate (PC), polypropylene (PP), and polyethylene terephthalate (polyester, or PET). The largest use is the PET beverage bottle. The same process of blow molding is used to make containers of glass.

When fibers of thermoplastics are stretched, the molecules become aligned and the fibers become very strong and resist being stretched any further. Once the fibers have been stretched, or *drawn*, they are strong enough for use in textiles and rope. Drawn fibers of nylon and polyesters are used for their strength.

Sheets of polymethylmethacrylate (PMMA), polycarbonate (PC), polyethylene (PE), polystyrene (PS), nylon, acrylics, and other polymers are produced by rolling. The process is similar to the rolling of metals.

Sheets of thermoplastics are *thermoformed*. This involves heating them enough to soften them and form them by drawing them into a mold with vacuum (Fig. 13.5).

Compression molding is a widely used molding process for thermosetting plastics. This consists of loading a precise charge in the bottom half of a heated mold, closing the mold halves to compress the charge, forcing it to flow and conform to the shape of the cavity, and heating the charge to polymerize it into a solidified part (Fig. 13.6).

Foams may be created by mechanical stirring or by gases formed by a chemical reaction that forms a gas or releases dissolved gases. Polystyrene foam is formed by the release of dissolved pentane on heating.

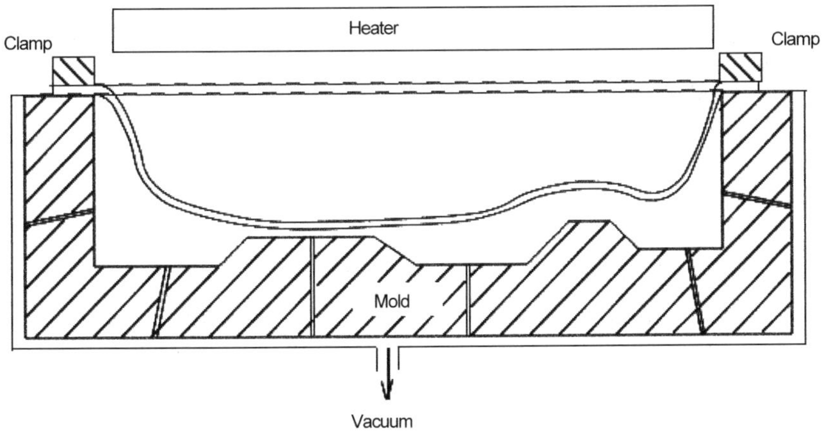

Fig. 13.5 Thermoforming of a thermoplastic sheet. Source: Ref 13.1

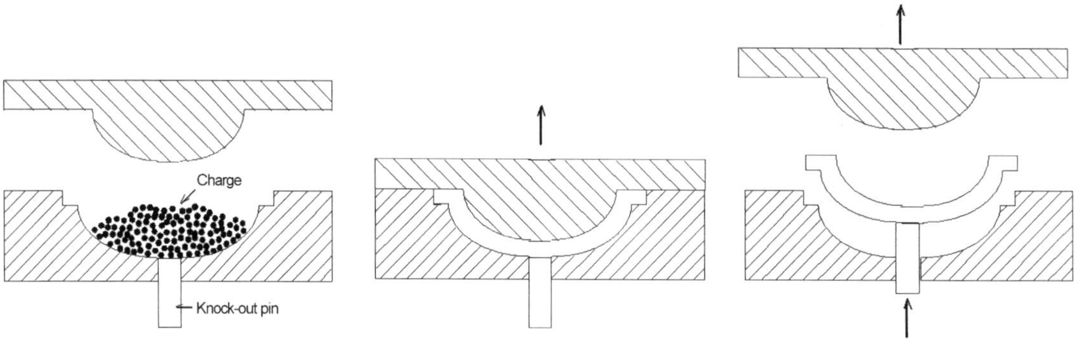

Fig. 13.6 Compression molding. Source: Ref 13.1

Polyurethane foam is formed by the release of water vapor produced as a byproduct of the polymerization reaction. In all cases the foam is formed into the desired shape by molding.

Powder Processing

Many products are made by pressing and sintering powders. Most ceramics are consolidated this way. Refractory oxides ceramics cannot be fabricated by melting and freezing. Sintering also is used to produce parts of metals that are difficult to melt. Examples include carbide tools and tungsten for lamp filaments. Mixed powders are sintered to make composites that are not otherwise possible, such as friction materials for brakes and clutches. Porous parts for filters or oil-less bearings are made by incomplete sintering. Teflon cannot be melted without decomposing so it also is processed as a powder. Pharmaceutical pills are made from powder. Powder processing is a simple and inexpensive way of fabricating large numbers of parts.

Figure 13.7 illustrates schematically how a part is pressed from powder. The process is highly automated with many parts being pressed per second.

There are limitations on the shape of the die: it must be prismatic so the compact can be ejected. The ratio of height-to-diameter must not be too great. Otherwise, friction on the sidewalls of the die will not allow sufficient compaction pressure in the center.

The *sintering* of pressed powders at elevated temperature bonds together the small powder particles without melting them. The driving force for sintering is the reduction surface area and the associated energy. In the initial stages of sintering, adjacent particles adhere to each other and form a neck where they are in contact. Figure 13.8 is a micrograph of such a neck formed between two nickel spheres.

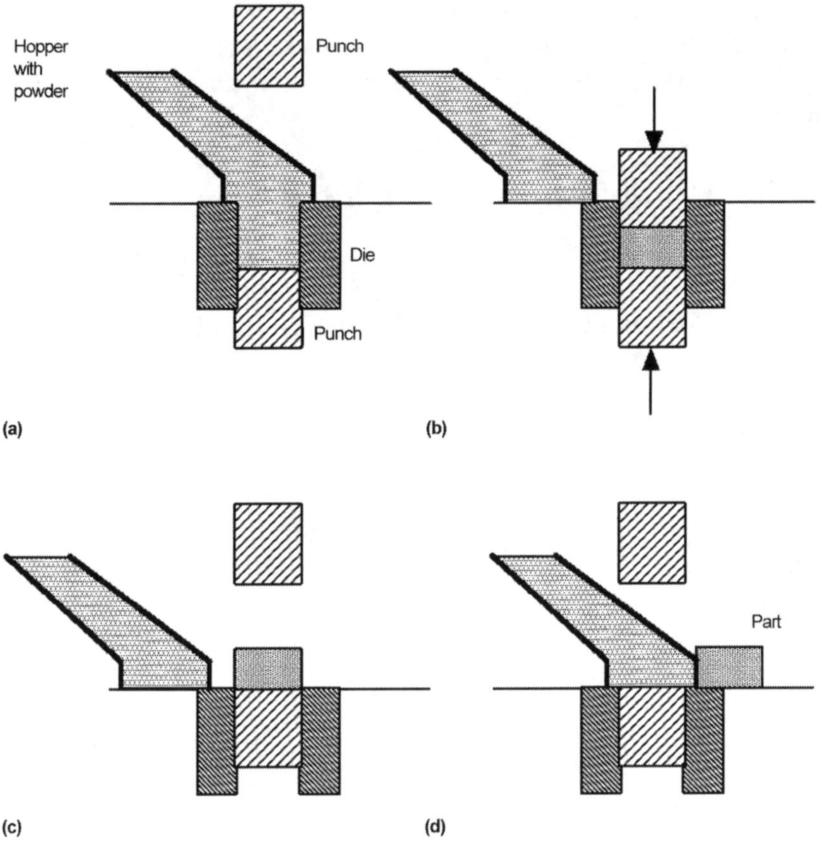

Fig. 13.7 Powder compaction. (a) A die is filled with powder. (b) The powder is compacted. (c) The compact is ejected. (d) The die is refilled as the part is pushed away. Source: Ref 13.1

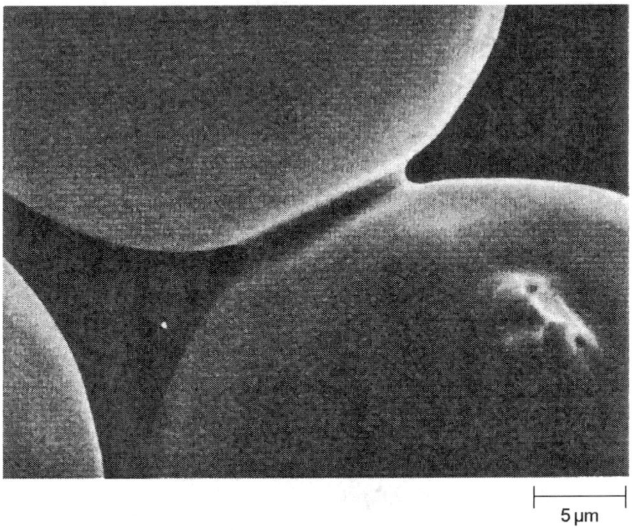

Fig. 13.8 Partially sintered nickel spheres. Source: Ref 13.3

As sintering progresses, the amount of surface area and the energy associated with the surfaces decrease. The density increases and the porosity decreases. The decrease in porosity is accompanied by an increase in strength.

Modern Manufacturing Techniques

The intricate circuits formed on semiconductor chips are patterned by *photolithography*. A photosensitive material (*photoresist*) is applied to a semiconductor surface and baked at a low temperature. The surface then is exposed to intense ultraviolet light, which makes the exposed regions of the photoresist become soluble in a developer. After these areas are removed, the surface is baked at a somewhat higher temperature to harden the remaining photoresist. The exposed regions then are acid etched.

The etched regions then may be treated differently from the rest of the surface. They may be *doped* by exposure to plasma containing *n*- or *p*-type impurities, or they may be plated to form conducting circuits. Integrated circuits are composed of many overlapping layers (Fig. 13.9), each defined by photolithography. *Dopants* are diffused into the substrate; additional ions are implanted into some layers. Metal and polycrystalline silicon are deposited on others to form conductive circuits.

Transistors can be formed where there are *n*-*p*-*n* or *p*-*n*-*p* diffusion layers. Resistors are formed by creating long narrow strips of a resistive conductor. Capacitors are formed by parallel conductors separated by insulating material.

Fig. 13.9 Integrated circuit with many layers. Source: Ref 13.4

NOTES OF INTEREST

Aluminum foil is produced by simultaneously rolling two sheets of aluminum. When the sheets are separated, the sides that were in contact with each other have a matte finish, the roughness of which reflects the grain size. The other sides are shiny because of sliding on the polished steel rolls.

Many consumer goods are shrink-wrapped. Shrink-wrapping uses films of PVC or low-density PE that have been stretched biaxially during manufacture. On heating they tend to contract, so if wrapped around an object and heated with hot air, they shrink to form a tight package. The same phenomenon is used to make tubes that serve as electrical insulators. Prestretched tubes placed over a wire joint decrease their diameters, forming a tight insulation when heated.

Three-dimensional parts can be made by photostereolithography or 3-D printing using a liquid that polymerizes on exposure to ultraviolet light. A substrate platform is lowered into a bath and a desired portion of the surface of the liquid is exposed to a beam of ultraviolet light, causing it to polymerize. A computer data file controls the area to be exposed to the scanning beam. After the surface is scanned, the substrate is lowered, allowing an additional thin layer of liquid to flow over the surface. The process is repeated until a three-dimensional structure is polymerized. Finally, the liquid that has not polymerized is drained off, leaving a solid part.

Dense sintered parts of inorganic materials can be formed by adding sinterable ceramics or metals to the liquid and firing the polymerized shape to drive off the organics. Typically the liquid may contain 40 to 70% sinterable ceramic and/or metallic particles, 10 to 35 % photocurable monomer, and 1 to 10% photoinitiators and dispersants.

This process has been used to make scaffolds for bone growth that can be implanted where bones are missing, for internal cores to allow air passages in cast turbine blades, and other parts that require intricate ceramic shapes.

REFERENCES

13.1 W.F. Hosford, *Materials for Engineers*, Cambridge University Press, 2008

13.2 W.F. Hosford and R.M. Caddell, *Metal Forming: Mechanics and Metallurgy*, 4th ed., Cambridge University Press, 2010

13.3 R.M. German, *Powder Metallurgy Science*, Metal Powder Industries Federation, Princeton, NJ, 1984

13.4 http://en.wikipedia.org/wiki/Integrated_circuit, accessed 2007

CHAPTER 14

Recycling

Metals

More than 80% of steel is made from recycled material. Steel and iron can easily be separated magnetically from other scrap. The scrap is either remelted in an electric arc furnace or added to pig iron in a basic oxygen furnace. All grades of steel can be recycled because most alloying elements are oxidized during processing. Tin and copper are the exceptions, and there is concern in the steel industry about the gradual buildup of these *tramp* elements in steel.

Recycling of nonferrous metals involves two types of scrap. One is the *offal*, or waste, from a manufacturing plant. It consists of trimmings from sheets, parts rejected for failure to meet standards, and other sources where the composition of the material is well known. Because the composition of new scrap metal is well known, it can be remelted to make more of the same alloy. This is much more valuable than *old* scrap of unknown composition. For most nonferrous metals, alloying elements cannot be removed during melting. Therefore, old scrap must be used to produce alloys with less critical compositions.

A substantial amount of aluminum is recycled. It was estimated that in 1998, 40% of the world's aluminum was produced by recycling. Aluminum scrap is shredded, and any lacquer is removed from cans before remelting. Remelting aluminum requires only 5% of the energy that is required to produce virgin aluminum from bauxite, and emits only 5% of the CO_2. The energy saving is 14 kwh/kg. In the United States, nearly half of all aluminum beverage cans are recycled; the rest go into landfills. The recycle rate in Europe is approximately 40%. Almost 90% of the aluminum cans are recycled in Norway, Sweden, and Switzerland. The rate of recycling in Brazil and Japan is over 90%.

Recycled copper accounts for approximately 40% of the U.S. consumption of copper. Recycling uses approximately 15% of the amount of energy that is used to produce copper from ore. In general, the value of recycled copper is approximately 90% that of virgin copper. The primary sources are electrical wiring, plumbing pipes, and fixtures. Control of the scrap is important. To produce new wire, only high-conductivity (highly pure) scrap can be used. Some impurities may cause *hot shortness* (cracking during hot working). Copper and copper alloy scrap must be carefully segregated. Pure copper, not contaminated by other metals, can be used to produce high-quality products. Alloy scrap that is segregated can be used for products of similar composition. Copper that is mixed with other metals, perhaps by having been tinned or soldered, or alloys to which lead has been added for machinability, can be used in alloys—such as bronze—that contain these metals.

At present, approximately 70% of the zinc produced originates from mined ores, and 30% from recycled zinc. The level of recycling is increasing in step with progress in zinc production technology and zinc recycling technology.

The supply of zinc-coated steel scrap increases annually as more auto body steel is galvanized. Approximately half of the world's steel is produced in electric arc furnaces. In the process, flue dust with high zinc contents is treated to recover zinc.

Lead battery plates are melted, refined, and recycled. The plastic case is shredded and recycled. The U.S. Environmental Protection Agency (EPA) estimates that approximately 80% of all lead-acid car batteries currently are recycled. Most battery recycling facilities will accept lead-acid batteries.

Different solder alloys contain different combinations of metal elements. The most common metals used in lead-free solder are tin, copper, silver, and sometimes zinc-antimony and bismuth. Recycling these materials, the user can reduce the environmental impact and save reprocessing and disposal costs.

Photographic film, plates, and paper; electrical and electronic parts; jewelry; and spent catalysts are sources of silver for recycling. It is estimated that 26% of the annual usage of silver comes from recycling.

Plastics

Thermoplastics account for 80% of all plastics and these can be recycled. Heating of thermosetting polymers will cause them to decompose without melting, so they can be used only as landfill. The American Plastics Council has established a system of symbols (Fig. 14.1) to identify several of the more common plastics for recycling.

Resins that already have been recycled are identified by an *R* before the previous designation. For example, RHDPE means recycled high-density polyethylene.

Recycling polyethylene uses only one-third of the energy that is used to produce new polyethylene. According to the EPA, recycling of 0.45 kg (1

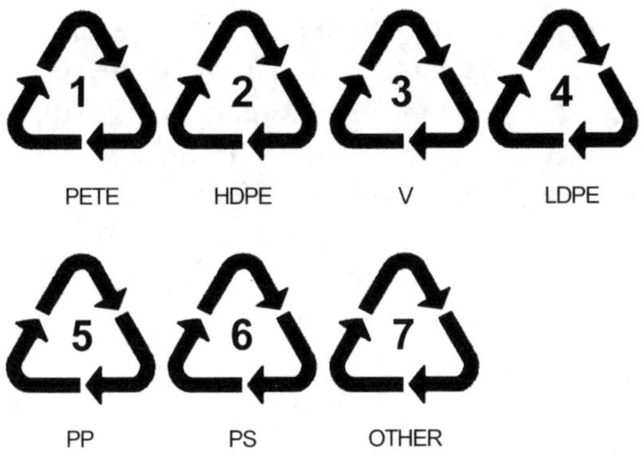

Fig. 14.1 Recycling symbols for polymers. (1) Polyethylene terephthalate, also indicated by PET and called polyester. (2) High-density polyethylene (also PE-HD). (3) Polyvinyl chloride, or PVC. (4) Low-density polyethylene (PE-LD), or LLDPE for very-low-density polyethylene. (5) Polypropylene. (6) Polystyrene. (7) Other resins or a combination of resins.

lb) of polyethylene terephthalate (PET) saves approximately 12.5 MJ of energy. Generally, recycled polymers are less desirable to manufacturers than new resins and are used in less critical applications. For example, virgin PET is used for bottles, and recycled PET is used mainly for fibers for carpets and fiberfill. New high-density polyethylene (HDPE) also is used for bottles, but recycled HDPE is used for drainage pipes, laundry detergent bottles, floor tile, and picnic furniture. New polyvinyl chloride (PVC) finds uses as squeezable bottles, frozen food bags, clothing, and carpets, while recycled PVC is used for floor tile, garbage can liners, furniture, trash cans, and lumber. Polypropylene (PP) is used in food and medicine packaging, but recycled PP is used for brooms, battery cases, ice scrapers, trays, and pallets. Virgin polystyrene (PS) is used for rigid and foamed products including coffee cups; among the recycled applications are foam packaging, thermal insulation, egg cartons, carryout containers, and rulers.

Rubber

Because most rubber products have been cross-linked by vulcanization, recycled rubber cannot be molded into new shapes. However, shredded rubber often is added to concrete or used in mats.

Glass and Paper

It is estimated that recycling of glass saves 50% of the energy that is used to make new glass from sand and limestone, and generates 20% less air pollution and 50% less water pollution. Recycling of paper uses only 40% of the energy that is used to make paper from wood and causes only 5% as much air pollution.

Many products that claim to be biodegradable should more properly be called oxi-degradable. Plastarch material (PSMpolyactide) and polyactide (PLA) do degrade by composting. Products made from petrochemical compounds generally do not biodegrade. Tree leaves are biodegradable. They are made in the spring, used by the plant during the summer, drop to the ground in autumn, and are assimilated into the soil. There are no microorganisms that can similarly break down most man-made products.

NOTE OF INTEREST

Recycling has been practiced throughout human history. Even before the industrial revolution, scrap bronze, tin, and other precious metals were collected and melted down for reuse because recycled metals were cheaper than virgin metals. The shortage of materials during World War II greatly encouraged recycling. Governments in every country involved in the war urged citizens to donate metals. Recycling programs established during the war were continued in some countries, such as Japan, after the war ended.

APPENDIX I

Greek Alphabet

Name	Greek use or letter	Meaning	Name	Greek use or letter	Meaning
alpha	Α		nu	Ν	
	α	angle		ν	
beta	Β		xi	Ξ	
	β	angle		ξ	
gamma	Γ		omnicron	Ο	
	γ	angle, shear strain, surface energy		ο	
delta	Δ	difference	pi	Π	multiplicative series
	δ	difference		π	3.14159...
epsilon	Ε		rho	Ρ	
	ε	true strain		ρ	
zeta	Ζ		sigma	Σ	summation
	ζ			σ	stress, conductivity
eta	Η		tau	Τ	
	η	viscosity, efficiency		τ	shear stress
theta	Θ		upsilon	Υ	
	θ	angle		υ	
iota	Ι		phi	Φ	
	ι			ϕ	angle
kappa	Κ		chi	Χ	
	κ			χ	
lambda	Λ		psi	Ψ	
	λ			ψ	pi
mu	Μ		omega	Ω	ohm, the end
	μ			ω	angular frequency

APPENDIX II

SI Unit System

The short forms of SI units are symbols—not abbreviations. They never end with a period. For example, 20 mm and 10 kg, not 20 mm. and 10 kg.

There should be a space between the quantity and the SI symbol. For example, 250 kg, not 250kg.

Symbols always are singular. For example, 100 kg, not 100 kgs. However, it is correct to write the unit name plural. For example, 100 kilograms.

The symbol for a compound unit is indicated by a slash or negative exponent. For example, km/h or km · h^{-1}, not kmph (using *p* as a symbol for per). It is acceptable to *say* "kilometers per hour."

Base Units

Quantity	Unit	Symbol
Length	meter	m
Mass	kilogram	kg
Time	second	s
Electric current	ampere	A
Temperature	Kelvin	K
Amount of substance	mole	mol
Luminous intensity	candela	cd

Derived Units

Quantity	Name	Symbol
Frequency	Hertz	hz
Force	Newton	N
Pressure, stress	Pascal	Pa
Energy, work	Joule	J
Power	Watt	W
Electric charge	Coulomb	C
Electric potential	Volt	V
Capacitance	Farad	F
Electric resistance	Ohm	Ω
Magnetic flux	Weber	Wb
Celsius temperature	Degree Celsius	°C

Standard Prefixes

Name	Symbol	Quantity
femto	f	10^{-15}
pico	p	10^{-12}
nano	n	10^{-9}
micro	μ	10^{-6}
milli	m	10^{-3}
kilo	k	10^{3}
mega	M	10^{6}
giga	G	10^{9}
tera	T	10^{12}
peta	P	10^{15}

APPENDIX III

Conversions and Constants

Conversions

1 amu = 1.66×10^{-24} g
1 angstrom (Å) = 10 nm = 10^{-10} m
1 calorie = 4.18 J
1 erg = 10^{-7} J
1 eV = 1.60×10^{-19} J
1 g = 2.20×10^{-3} lbm
1 inch = 2.54 cm
1 newton = 0.224 lbf
1 pascal = 0.145×10^{3} psi
1 poise = 0.1 Pa · s

Constants

Measure	Symbol	Value
Acceleration of gravity	g	9.80 m^2/s
Avogadro's number	N_o	6.022×10^{23} $mole^{-1}$
Boltzmann constant	k	86.1×10^{-6} eV/K = 13.8×10^{-24} J/K
Charge of an electron	q	1.60×10^{-19} C
Gas constant	R	8.31 J/mol × K
Planck's constant	h	6.62×10^{-34} J · s
Velocity of light	c	2.99×10^{8} m/s
Permittivity of free space	ε_o	8.85×10^{-12} farad/m
Atomic mass unit	amu	1.66×10^{-24} g

APPENDIX **IV**

Properties of Elements

Element	Symbol	Atomic number	Atomic weight, amu	Melting temp., °C (°F)	Density, Mg/m³	Atomic radius, nm	Crystal structure(a)
Hydrogen	H	1	1	–259 (–434)	...	0.46	...
Helium	He	2	4	–272 (–457)	...	0.176	...
Lithium	Li	3	7	181 (357)	0.534	0.152	bcc
Beryllium	Be	4	9	1290 (2354)	1.85	0.114	hcp
Boron	B	5	10.8	2300 (4172)	2.3	0.046	...
Carbon	C	6	12	>3500 (>6332)	2.25	0.077	hcp
Nitrogen	N	7	14	–210 (–346)	...	0.071	...
Oxygen	O	8	16	–218 (–360)	...	0.060	...
Fluorine	F	9	19	–220 (–364)	...	0.06	...
Sodium	Na	11	23	98 (208)	0.97	0.186	bcc
Magnesium	Mg	12	24.3	649 (1200)	1.74	0.161	hcp
Aluminum	Al	13	27	660 (1220)	2.70	0.143	fcc
Silicon	Si	14	28	1410 (2570)	2.33	0.1176	dia cub
Phosphorus	P	15	31	44 (111)	1.8	0.11	...
Sulfur	S	16	32	113 (235)	2.07	0.106	...
Chlorine	Cl	17	35.5	–101 (–149)	...	0.101	...
Potassium	K	19	39	63 (145)	0.86	0.231	bcc
Calcium	Ca	20	40	839 (1542)	1.55	0.197	fcc
Titanium	Ti	21	47.9	1668 (3034)	4.51	0.146	hcp
Vanadium	V	23	51	1900 (3452)	6.1	0.132	bcc
Chromium	Cr	24	52	1875 (3407)	7.19	0.125	bcc
Manganese	Mn	25	55	1244 (2271)	7.47	0.112	bcc
Iron	Fe	26	55.8	1538 (2800)	7.87	0.124	bcc
Cobalt	Co	27	58.9	1495 (2723)	8.83	0.125	hcp
Nickel	Ni	28	58.7	1455 (2651)	8.90	0.125	fcc
Copper	Cu	29	63.5	1084 (1983)	8.93	0.128	fcc
Zinc	Zn	30	65.4	420 (788)	7.13	0.139	hcp
Germanium	Ge	31	72.6	937 (1718)	5.32	0.1224	dia cub
Arsenic	As	32	74.9	816 (1500)	5.78	0.125	...
Zirconium	Zr	40	91.2	1852 (3365)	6.49	0.158	hcp
Niobium	Nb	41	92.9	2468 (4474)	8.57	0.143	bcc
Molybdenum	Mo	42	96	2610 (4730)	10.22	0.136	bcc
Silver	Ag	47	107.9	962 (1763)	10.5	0.1444	fcc
Tin	Sn	50	118.7	232 (449)	7.17	0.151	bct
Antimony	Sb	51	121.7	631 (1167)	6.7	0.1452	...
Iodine	I	53	126.9	114 (237)	4.93	0.135	...
Cesium	Cs	55	132.9	29 (84)	1.9	0.265	bcc
Tungsten	W	74	144	3410 (6170)	19.25	0.1367	bcc
Gold	Au	79	197	1064 (1947)	19.3	0.144	fcc
Mercury	Hg	80	200.6	–39 (–38)	...	0.155	...
Lead	Pb	82	207.2	327 (620)	11.38	0.175	fcc
Uranium	U	92	238	1133 (2071)	19.05	0.138	...

(a) bcc, body-centered cubic; fcc, face-centered cubic; hcp, hexagonal close-packed; dia cub, diamond cubic; bct, body-centered tetragonal

APPENDIX V

Properties of Materials

Material	Density, Mg/m^3	Thermal conductivity, W/m · °C	Linear expansion/°C	Electrical resistivity, ohm · m	Young's modulus, GPa
Aluminum	2.7	22	22×10^{-6}	29×10^{-9}	70
Brass (30% Zn)	8.5	12	20	62	110
Bronze (5% Zn)	8.8	8	18	100	110
Cast iron	7.15	...	10	...	20
Copper	8.9	40	17	17	16
Diamond	3.52	200	1.1	10^{11}–10^{15}	1050
Magnesium	1.74	16	25	45	45
Silicon	2.33	86.7	2.8	10^{-9}	113
Silver	10.4	41	18	18	75
Steel (1020)	7.86	5	11.7	169	205
Alumina	3.8	2.9	9	$>10^{12}$	50
Concrete	2.4	0.01	13	...	14
Glass (plate)	2.5	0.075	9	10^{12}	70
Graphite	1.9	25–470	5	10^{-5}	7
Magnesia	3.6	2.1	9	10^{12}	205
Quartz	3.17	1.2	...	10^{12}	310
SiC	3.17	1.2	7	50	350
Melamine	1.5	0.030	27	10^{11}	9
Formaldehyde phenol	1.3	0.016	72	10^{10}	3.5
Formaldehyde urea	1.5	0.030	27	10^{11}	10.3
Formaldehyde rubber (vulcanized)	1.2	0.012	81	10^{12}	3.5
Polyethylene (high-density)	0.96	0.052	120	10^{12}–10^{16}	0.35–1.25
Polyethylene (low-density)	0.92	0.034	180	10^{13}–10^{16}	0.1–0.35
Polystyrene	1.05	0.008	63	10^{16}	2.8
Polyvinyl chloride	1.4	0.19	50	...	2.6
Polytetrafluoroethylene	2.2	0.020	100	10^{14}	0.35–0.75
Polymethyl methacrylate	1.2	0.020	90	10^{14}	3.5
Nylon	1.15	0.025	100	10^{12}	2.8

APPENDIX VI

Unit Abbreviations

Symbol	Unit
A	ampere
Å	angstrom
Btu	British Thermal Unit
C	coulomb
°C	degrees Celsius
cal	calorie
eV	electron volt
°F	degrees Fahrenheit
J	joule
K	degrees Kelvin
kg	kilogram
ksi	thousand pounds per square inch
lbf	pounds force
lbm	pounds mass
mm	millimeters
MPa	megapascals
N	Newtons
Nm	Newton-meters
Pa	pascals
W	watts
psi	pounds per square inch

Index

B

C

D

E

F

G

H

I

J

K

L

M

N

O

P

Q

R

S

T

U

V

W

Y

Z